Contents

Overview

For centuries, people have been trying to gather facts about the sky and beyond. People have always wondered about the mysteries of space.

In the 4th century B.C., Aristotle believed the sun revolved around the earth. People respected and believed his theories for centuries. Twenty centuries later, Copernicus put forth the theory that the earth revolved around the sun. He formed his theory based on naked-eye observations. His views were widely rejected. However, his thoughts made sense to some people and slowly, a scientific revolution began.

At the beginning of the 17th century A.D., Galileo built a telescope and began studying the skies more closely. He became very unpopular because he insisted that the earth revolved around the sun. Galileo studied the heavens with his telescope until he could explain his theories. However, the people of his time preferred to believe Aristotle's thoughts rather than Galileo's proofs. Even though his ideas were not accepted by the masses, some people could not reject the logic in what he was presenting.

Sir Isaac Newton was one of the people who took a serious look at Galileo's ideas. At the end of the 17th century, Newton was able to mathematically prove that Galileo was right. Unlike Galileo, Newton was not unpopular because of his beliefs. People did not reject his ideas. He became a popular hero. From Newton's time on, accepted fact has been that the earth and the other planets revolve around the sun.

By the time the 20th century rolled around, human beings wanted to do more than just look up at outer space. People wanted to go there. And so it has come to be! This book presents the fascinating story of the human experience in the sky and in outer space.

As you read this story, you will also learn how to get the most out of reading a textbook. Along with reading the text, you will complete some activities. These activities are explained below.

Setting the Stage
Setting the Stage is designed to help you get some quick ideas about the lesson you are going to read. You will quickly glance at the first two paragraphs and the last two paragraphs. Then, you will write a few words that give you a general idea what the lesson will be about.

AGS The History of Space Travel

Reading in the Content Area

by
Sue Boulais

American Guidance Service, Inc.
4201 Woodland Road
Circle Pines, MN 55014-1796
1-800-328-2560

Photo credits:
cover—(Robert H. Goddard standing beside the first liquid fuel rocket in 1926, shuttle on launch pad) Photri;
p. 7—NASA; p. 8—Culver Pictures, Inc./SuperStock; pp. 14, 26—NASA; p. 32—Culver Pictures, Inc./SuperStock;
p. 38—NASA; p. 44—Photri-Microstock; p. 50-Photri; pp. 55, 56—NASA; p. 62-Photri; pp. 68, 73, 74—NASA;
p. 80—JSC NASA; p. 86 (top right)—Bob Daemmrich/Stock Boston, (top left)—Richard Pasley/Stock Boston,
(bottom)—Jeff Dunn/Stock Boston

Printed in the United States of America

ISBN 0-7854-2432-6

Order Number: 91590

A 0 9 8 7 6 5 4 5 3

Discussing the Background

Studies show that using information you already know can help you understand new information you read. To give you the best possible chance to understand what you are reading, you will be asked to discuss background information. After the first lesson, this information will come from the previous lesson. This method works well because each lesson follows in order through the history of space travel.

Words to Know

As you know, reading is difficult when you do not understand the words you are reading. Although you will work with the vocabulary words after you read, you should read the words in the Words to Know box to become familiar with them. As you read, if you come to a boldface word you do not know, you can look it up in the glossary.

Finding the Main Idea

Every reading selection has a main idea. Often, a group of paragraphs within an article or essay are related and have a main idea. You will be asked to highlight or circle the main idea in each paragraph. The main idea does not have to be complicated. It can be a single word, a phrase, a sentence, or more than one sentence.

When you are finished marking the main ideas, take a moment to read just the sentences and phrases you have marked. You will find you have a good overview of the lesson.

Making a Timeline

The purpose of the timeline is to see how different events relate to each other. You will be asked to place important dates from each lesson on a timeline. Sometimes, you will be asked to place dates from lessons you have already read on the timeline.

Using Context Clues

Context clues are one of the main ways readers have to figure out the meanings of unknown words. The words and sentences around an unknown word can give a lot of clues to the meaning of the word. Often, you will not learn exactly what the word means, but you will have a fairly good idea. In this exercise, you will be asked to give the meaning of a word and list a context clue for each word.

Overview

Reading for Details
Reading for Details teaches you to find details that support main ideas. You will be asked to fill in the information in webs. These diagrams are a way to visually organize information.

In the early lessons, you will work with one paragraph at a time. The main idea will be the same one you highlighted or circled. Later on, you will be asked to complete main idea/detail webs on groups of paragraphs. To do this, look at all the main ideas involved and create a main idea statement that includes all of them.

Once you have written a main idea in a web, write three details to support the main idea. In many cases, there are more than three possible details. You can choose to write only three or to combine two or more of them into one space.

These main idea/detail webs can be used to outline an entire article. This task could be completed by using a single web for each paragraph or by combining paragraphs on webs.

Other Activities
Some units and lessons have specific activities presented on the last page. These activities specifically relate to the unit or lesson topics.

Getting the Main Idea

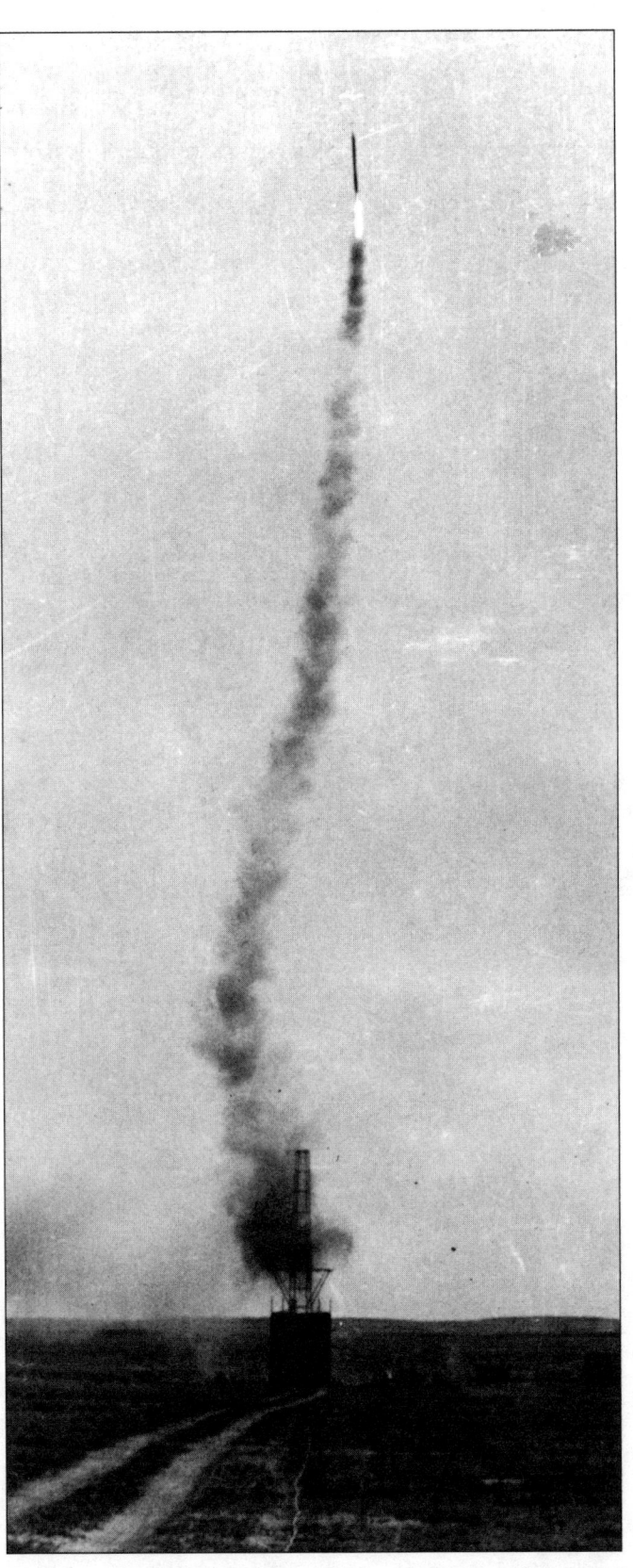

One of Robert Goddard's rockets flies over Roswell, New Mexico, on August 26, 1937.

Every written selection, from paragraph to article, has a main idea. The main idea tells the selection's topic and what the selection says about that topic.

Often, the main idea of a selection will be the first or last sentence in the selection. This is called an *explicit* main idea, because the main idea is stated. Other selections may contain sentences that only hint about the main idea, or are *implicit*. Knowing a selection's main idea helps you to better understand what the selection says.

Two tools that will help you find the main idea are *context* and *details.* Context is the way words, sentences, and paragraphs work together to provide meaning. Details are bits of information that add to your understanding of a subject.

You will learn more about the main idea, context, and details in the next four lessons.

The Wright brothers test out *Flyer* at Kitty Hawk, N.C. *Flyer* was the world's first heavier-than-air flying vehicle that was pilot-controlled.

Orville and Wilbur Wright and *Flyer*

A. Setting the Stage

Read the first two and last two paragraphs of this lesson. Write four nouns or phrases that appear to be key ideas.

1.	2.	3.	4.

B. Discussing the Background

As a group, answer and discuss these questions: Why are Orville and Wilbur Wright famous? How long before you were born did the first airplane fly?

Words to Know

contraption

convince

technology

adapt

witness

thrust

1 The day was December 17, 1903. Orville and Wilbur Wright stood on a windy beach at Kitty Hawk, North Carolina. Near them was their latest flying machine. Two 40-foot wooden wings covered with cloth were fastened together with wire, one above the other. The bottom wing sported a cradle and a lever. Next to these was a small engine connected to two propellers by a bicycle chain. The brothers called this **contraption** a biplane. They claimed it would fly.

2 Their claim did not create much interest. Three days before, Wilbur tried to get the thing in the air and ran into a sand dune. Besides, inventors had been trying—and failing—to fly machines for hundreds of years. Most people believed that humans were not meant to fly.

3 But the Wright brothers weren't most people. They were **convinced** that people could make a flying machine. They were also convinced that their biplane, *Flyer*, would be the first successful flying machine. The brothers believed that flight control was the answer. They saw this control as the most important quality of a flying machine.

4 Finding a way to control a machine's flight would give flying a purpose. Control meant having a pilot on, or in, the machine. The pilot had to be able to choose when to go up in the air. Likewise, the pilot needed to control the landing. Flying machines of the time—hot-air balloons, kites, and gliders—could carry a person. But the flight was controlled by the wind, not by the passenger.

Even though they were wind-controlled, experiments with kites and gliders helped the Wrights build *Flyer.* In fact, *Flyer* was the latest thing in flying **technology.** It combined the newest kind of kite, the newest kind of glider steering, and the newest kind of engine.

5

Flyer's wings were really a box kite. The box kite was invented in 1893 by Australian Lawrence Hargrave. The shape of the kite made it steady in the air. The brothers' experiments showed that box kites would be easier to control in the air.

6

The brothers developed *Flyer's* steering system from Otto Lilienthal's work during the 1890s. Lilienthal was a glider pilot. He gained some control over his glider by turning his body while flying. The Wright brothers **adapted** Lilienthal's idea with their hip cradle. Between cradle and wing tips they attached wires, four in all. The pilot lay flat in the cradle and turned the cradle with his body. The turning cradle pulled the wires to move the wings. *Flyer's* wing tips could be turned left, right, up, and down.

7

"With a short dash down the runway, the machine lifted into the air and was flying . . . it was a real flight at last and not a glide."
—Orville Wright

The Wrights had to be creative to get *Flyer* off the ground. They knew a gasoline powered engine had recently been developed. They used the same method to create a 12-horsepower engine for *Flyer.* Most other inventors of the time were trying to use steam engines to power their crafts. The Wrights felt that steam engines were too heavy.

8

The brothers also put in a lever to point *Flyer's* nose up or down. Wilbur had problems with the lever. Because of the lever, *Flyer* crashed on its first try. Wilbur pulled *Flyer's* nose too high. The brothers repaired the slight damage and planned to move the lever less next time. On December 17, 1903, *Flyer* was ready to fly under its own power, without the wind.

9

Orville climbed onto *Flyer's* cradle and started the engine. The propellers spun, pushing the craft along its wooden rail. Wilbur ran beside, holding one wing level. Orville pulled the lever, not too much this time. *Flyer* left its rail, rose in the air, and flew for 12 seconds and 120 feet. The brothers tried it three more times. *Flyer* got off the ground and stayed off each time. Orville made the first flight that day, but Wilbur made the longest: 852 feet in 59 seconds.

10

Orville and Wilbur Wright and *Flyer*

Only five people were at Kitty Hawk to **witness** the flight that gave humans wings. Most people saw only wooden wings, a cradle with wires, a tiny engine, and a lever. They could not see ahead to planes that would fly across countrysides, between cities, across countries, and across oceans. People were not aware that this flight would lead to planes that could fly faster than sound to the edge of Earth's atmosphere. The world did not realize that this contraption would lead to a flying craft that would take people past the atmosphere and into space.

11

Scientists wanted to go faster than sound, get past the pull of Earth's gravity and blast into space. This plan required incredible amounts of **thrust**. To get that pushing power, the airplane was coupled with an invention that other engineers and designers were working on. That invention was the rocket.

12

C. Finding the Main Idea

Highlight or circle the main idea in each paragraph. Remember that the main idea of a paragraph is often the first or last sentence in that paragraph. Other times, the main idea has to be pieced together from more than one sentence.

D. Making a Timeline

Place two important dates from this lesson on the timeline. Write two or three words to identify the importance of each date.

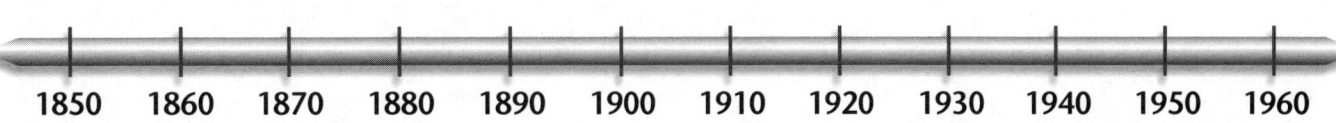

| 1850 | 1860 | 1870 | 1880 | 1890 | 1900 | 1910 | 1920 | 1930 | 1940 | 1950 | 1960 |

E. Using Context Clues

Find the following four vocabulary words in the lesson. The words and sentences around each word give a clue to its meaning. Use these clues to write a meaning for each word. For each vocabulary word, describe one clue that you used.

1. contraption

Meaning: _____

_____

_____

Clue: _____

_____

2. technology

Meaning: _____

_____

_____

Clue: _____

_____

3. adapted

Meaning: _____

_____

_____

Clue: _____

_____

4. thrust

Meaning: _____

_____

_____

Clue: _____

_____

F. Reading for Details

Reread the two paragraphs identified below. Then, complete the main idea/detail webs.

Paragraph # 7 Main Idea: _____	Supporting Detail #1
	Supporting Detail #2
	Supporting Detail #3

Paragraph # 9 Main Idea: _____	Supporting Detail #1
	Supporting Detail #2
	Supporting Detail #3

Orville and Wilbur Wright and *Flyer*

G. Hunting for Paragraphs

Find the paragraph that answers each question below. Write the paragraph numbers on the blank lines.

1. The Wright Brothers' flying experiment was more significant than all previous flying experiments. Which paragraph explains why? _____

2. Which paragraph explains why there were not very many people at the first flight? _____

H. Going Beyond the Main Idea

One of the pleasures of reading is expanding what you read in different questions. Go beyond the main ideas in the stories to answer the following questions.

1. How might the observers at Kitty Hawk have been different if people could have seen into the future?

   _____

   _____

   _____

2. Even though there were only two brothers, explain how the Wrights' *Flyer* was the result of a large team effort.

   _____

   _____

   _____

Robert Goddard's early rockets led to the development of larger rockets like this Saturn V rocket used to launch *Apollo 16.*

Robert Goddard and Rockets

A. Setting the Stage

Read the first two and last two paragraphs of this lesson. Write four nouns or phrases that appear to be key ideas.

1.	2.	3.	4.

B. Discussing the Background

Reread the last paragraph of Lesson 1. As a group, discuss the relationship between Lesson 1 and Lesson 2.

Words to Know
device
accurate
evolve
impress
substitution
theorize
data
propellant
patented
streamlined

1 Chinese rockets flared into history in the early 1200s. The people of China also invented kites, history's first flying machine.

2 Chinese armies fired these "arrows of flying fire" upon invading Mongol troops. These first rockets were simply arrows with tubes of black powder tied to them. Use of these "fire sticks" spread quickly. There were soon used in the Middle East, where Arabs and Europeans were fighting the Crusades. French Crusaders took the knowledge of rockets back to France with them. In 1429, Joan of Arc's French troops used the exploding **devices** in the Battle of Orleans.

3 These first rockets were unpredictable. Because of this problem, they were not widely used as weapons in Europe. When lit, they were just as likely to explode as fly. Their flight could not be controlled, so they were not very **accurate**. Guns and cannons became the accepted weapons. They were far more accurate and safer for the user.

4 For the next 350 years, rockets were used mostly for fireworks displays. In the late 1700s, British troops met them again in India. By then, they had **evolved** into small, metal-cased rockets. The British were **impressed** enough to reconsider rockets as weapons. British Colonel William Congreve improved the simple metal rockets. He turned them into more efficient weapons with exploding warheads up front. In the 1840s, another Englishman, William Hale, replaced the rocket's wooden tail with three fins. This **substitution** made rocket flight more accurate.

Scientists also took another look at rockets. They were part of the growing interest and experimentation in flight. By the 1890s, a few scientists were studying rockets and their flight possibilities.

5

Congreve's warhead and Hale's tail fins were minor changes. The first really important change in rockets came in a backyard in Auburn, Massachusetts. There, on March 16, 1926, American scientist Robert Goddard launched the first liquid-fueled rocket. The tiny 10-foot missile flew 40 feet high. It covered 184 feet in 2-1/2 seconds. Goddard used gasoline and oxygen for fuel.

6

"The first flight with a rocket using liquid propellants was made yesterday at Aunt Effie's farm . . . . It looked almost magical as it rose . . . as if it said, 'I've been here long enough; I think I'll be going somewhere else . . .' "
—Diary of Robert H. Goddard, March 17, 1926

Robert Goddard's success crowned 17 years of research and design. In 1909, he **theorized** that liquid fuel would work better than solid fuels. The solid fuels of the time burned quickly. They did not produce enough thrust to get a rocket high enough to be usable. He thought this fuel would give a rocket more power for a longer time.

7

Goddard worked to prove his theories with experiments and **data.** Liquid fuels, or **propellants**, needed a much more complicated rocket structure than was available. Much of Goddard's work between 1909 and 1926 was spent developing a rocket that could use liquid fuel.

8

Goddard's backyard launch attracted no more attention than the Wright brothers' first flight. The press and public showed little interest. Thus, the little rocket that opened the door to big changes went unnoticed.

9

During the 1920s and 1930s, Goddard built and launched even bigger rockets. Soon he was the leading rocket researcher in the United States. By the end of the 1930s, his rockets flew faster than the speed of sound (about 760 miles per hour). He developed and **patented** ideas that would be used later in bigger rockets such as liquid-fueled and self-cooled motors, guidance and control systems, fuel pumps, multi-stage rockets, and retrorockets for braking. But like Goddard's first launch, none of these got much notice.

10

Robert Goddard and Rockets

All attention was going to the airplane. The 1920s and 1930s were the Golden Age of airplane development. Gone were the wooden frames covered with cotton cloth. Every year, engineers designed safer, more powerful, more **streamlined** metal planes. Planes transported mail, goods, and people. Automatic pilots made flying easier and safer. Pressurized cabins made it comfortable for people to breathe in flight.

11

During these years, Richard Byrd flew across the North and South Poles. Charles Lindbergh flew across the Atlantic. Hugh Herndon and Clyde Pangborn flew across the Pacific. Amelia Earhart became the first woman to fly across the Atlantic alone. Wiley Post flew around the world.

12

World War II led to more design changes in airplanes. Bombers could carry twice as much and travel twice as far. Late in the war, the Germans perfected a rocket plane that flew combat missions. At the time of his death in 1945, Goddard was working on a U.S. combat rocket plane.

13

Rockets as a separate field of study seemed about to disappear. Suddenly, from Germany, streaked a deadly, destructive rocket. This rocket blasted the world awake to the possibilities of rocket technology. It also destroyed cities and killed thousands during the last months of the war. Its name was V-2.

14

C. Finding the Main Idea

Highlight or circle the main idea in each paragraph. Remember that the main idea of a paragraph is often the first or last sentence in that paragraph. Other times, the main idea has to be pieced together from more than one sentence.

D. Making a Timeline

Place three important dates from this lesson on the timeline. Also place one important date from Lesson 1. Write two or three words to identify the importance of each date.

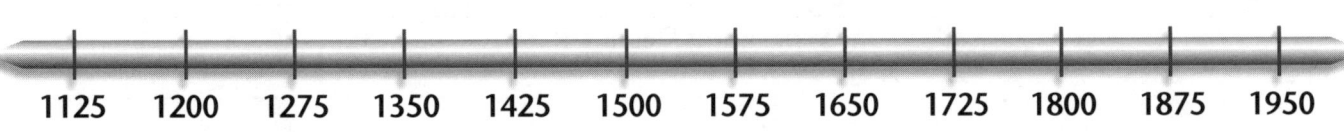

| 1125 | 1200 | 1275 | 1350 | 1425 | 1500 | 1575 | 1650 | 1725 | 1800 | 1875 | 1950 |

E. Using Context Clues

Find the following four vocabulary words in the lesson. The words and sentences around each word give a clue to its meaning. Use these clues to write a meaning for each word. For each vocabulary word, describe one clue that you used.

1. evolved

Meaning: _____

_____

_____

Clue: _____

_____

2. theorized

Meaning: _____

_____

_____

Clue: _____

_____

3. propellants

Meaning: _____

_____

_____

Clue: _____

_____

4. streamlined

Meaning: _____

_____

_____

Clue: _____

_____

F. Reading for Details

Reread the two paragraphs identified below. Then, complete the main idea/detail webs.

Paragraph # 6 Main Idea: _____ _____ _____	Supporting Detail #1
	Supporting Detail #2
	Supporting Detail #3

Paragraph # 11 Main Idea: _____ _____ _____	Supporting Detail #1
	Supporting Detail #2
	Supporting Detail #3

G. Hunting for Paragraphs

Find the paragraph that answers each question below. Write the paragraph numbers on the blank lines.

1. Which paragraph talks about a replacement that gave people more control over rockets?

   _____

2. Which paragraph talks about a time when rockets finally got the attention of the media?

   _____

H. Going Beyond the Main Idea

Go beyond the main ideas in the lesson to answer the following questions.

1. Given that few people paid attention to Goddard's work, why do you think he continued?

   _____

   _____

   _____

2. Paragraph 11 talks about wooden frames with cotton cloth. What does this have to do with rockets?

   _____

   _____

   _____

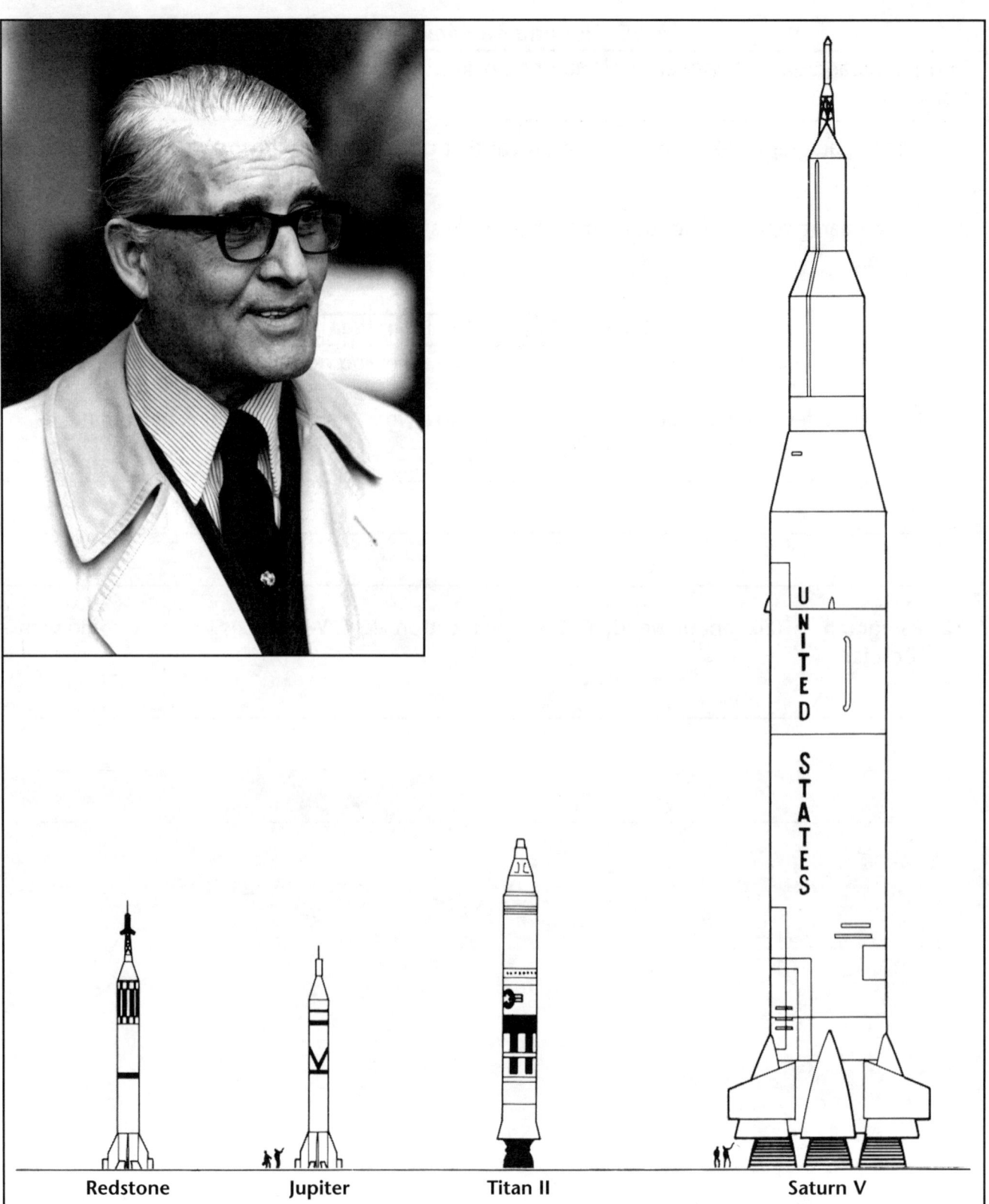

Top: Dr. Werner von Braun
Bottom: Von Braun's rockets

Redstone   Jupiter   Titan II   Saturn V

Oberth, Von Braun, and the V-2

Lesson 3

A. Setting the Stage

Read the first two and last two paragraphs of this lesson. Write four nouns or phrases that appear to be key ideas.

1.	2.	3.	4.

B. Discussing the Background

Reread the last paragraph of Lesson 2. As a group, discuss the relationship between Lesson 2 and Lesson 3.

Words to Know
annihilate
fine-tune
satellite
supersonic
sound barrier
initial

1 The V-2 was a powerful guided missile. The technology of the V-2 opened a new era of rocketry. The work of Hermann Oberth, Germany's "father of space," led to the development of the V-2. While Robert Goddard was working with rockets in the United States, Oberth was studying them in Germany. Neither man knew of the other or his work. Still, their ideas about liquid-fueled rockets were very similar. In 1923, Oberth published his ideas in *The Rocket into Interplanetary Space.* An intense interest in rockets grew throughout Germany.

2 In 1927, Oberth and other scientists formed the German Society for Space Travel. One of the earliest members was a young engineer named Werner von Braun.

3 Von Braun had dreamed of rockets and space travel for years. His interest began as a boy while reading Jules Verne's *From Earth to the Moon.* Von Braun also read Oberth's book. He wanted to work with Oberth. Von Braun got his wish. He and Oberth began developing and testing Europe's first liquid-fueled rockets. In 1930, they successfully launched a rocket and were on their way into space.

4 By the mid-1930s, however, the Nazi war machine changed everything. Oberth's and Von Braun's dreams of space travel were **annihilated.** Chancellor Adolf Hitler was gearing Germany up for war. The German military took over the men's rocket plans and research. Oberth was sent home to Romania (then Hungary) to teach. Von Braun was sent to work for the army. He was put in charge of creating rockets to be used as Hitler's Vengeance weapons, or *Vergetunswaffe.*

UNIT 1 *GETTING THE MAIN IDEA* 21

By the early 1940s, Von Braun's team developed a rocket that could carry a one-ton warhead. Oberth was called to help. He designed fuel pumps necessary for the huge rocket, now known as the V-2. Oberth completed his end of the job in 1943. His next assignment was to develop an antiaircraft missile, a rocket that could stop a plane or another missile like the V-2. Von Braun continued to **fine-tune** the V-2.

5

The first V-2 rocket to hit London exploded on September 8, 1944. It was 45 feet long and weighed more than 27,000 pounds. During the next seven months, Germany pounded European cities with the deadly guided missiles. More than a thousand hit on or near London alone. No defense against them was found before the war's end in 1945.

6

> "We [the V-2 team] are convinced that a complete mastery . . . of rockets will change conditions in the world . . . this . . . will apply both to the civilian and military aspects of their use."
> —Werner von Braun on his surrender to American troops, May 1945

During those same seven months, Von Braun looked for a way to escape from Germany. He could see Germany was about to lose the war. Hitler did not want his enemies to have the V-2 program. Von Braun was afraid the German police would kill him and his staff. He believed the V-2 plans and research would be destroyed to keep the program from the Allies. In May of 1945, Von Braun found a way to surrender to American troops.

7

During a secret effort called Operation Paperclip, more than one hundred German rocket specialists accompanied Von Braun to the U.S. So did one hundred unused V-2s captured by the Americans. Von Braun and his team wanted to share their rocket expertise with America. They were happy to get back to peace and on the road to space.

8

The Von Braun team and the V-2s went first to a U.S. Army base in New Mexico. There, the missiles were made into scientific research vehicles. The one-ton warheads were replaced with payloads carrying scientific equipment. From 1945 to 1952, the V-2s launched equipment that gathered information about the atmosphere. Collected data included temperatures, pressures, composition, densities, winds, magnetic fields, and radiation measurements.

9

Oberth, Von Braun, and the V-2

After Robert Goddard's death in August of 1945, Von Braun became the leading U.S. rocket researcher. Von Braun worked in New Mexico until 1950. Then the Army asked him to design a new, larger rocket. Von Braun patterned his new rocket, the Redstone, after the V-2. He also developed a plan called Project Orbiter: the Redstone would launch a **satellite!**

10

The dozen years after World War II also saw major changes in airplanes. German rocket planes appeared in the skies very close to the end of the war. U.S. engineers continued working towards **supersonic** flight machines that could fly faster than sound. On October 14, 1947, Air Force Captain Charles "Chuck" Yeager broke the **sound barrier** in his Bell X-1 rocket plane, *Glamorous Glennis*. From that **initial** breakthrough, the Air Force continued to build sleeker and faster supersonic rocket planes.

11

Jet engines came into their own, too. By 1957, the military used jet engines in fighter planes, bombers, and transports. Commercial passenger airlines were on the brink of using jet engines.

12

By 1957, humans had wings. Only a half-century earlier, people asked: Is it possible? Now people wondered: How fast? How far? How high? On October 4, 1957, a little satellite named *Sputnik I* answered.

13

C. Finding the Main Idea

Highlight or circle the main idea in each paragraph. Remember that the main idea of a paragraph is often the first or last sentence in that paragraph. Other times, the main idea has to be pieced together from more than one sentence.

D. Making a Timeline

Place three important dates from this lesson on the timeline. Also place one important date from Lesson 2. Write two or three words to identify the importance of each date.

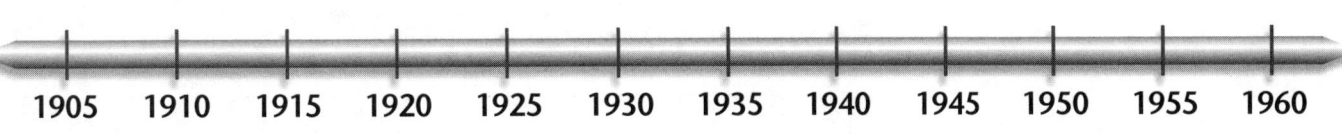

| 1905 | 1910 | 1915 | 1920 | 1925 | 1930 | 1935 | 1940 | 1945 | 1950 | 1955 | 1960 |

E. Using Context Clues

Find the following four vocabulary words in the lesson. The words and sentences around each word give a clue to its meaning. Use these clues to write a meaning for each word. For each vocabulary word, describe one clue that you used.

1. annihilated

Meaning: _____

_____

_____

Clue: _____

_____

2. fine-tune

Meaning: _____

_____

_____

Clue: _____

_____

3. supersonic

Meaning: _____

_____

_____

Clue: _____

_____

4. sound barrier

Meaning: _____

_____

_____

Clue: _____

_____

F. Reading for Details

Reread the two paragraphs identified below. Then, complete the main idea/detail webs.

Paragraph # 4 Main Idea: _____ _____	Supporting Detail #1
	Supporting Detail #2
	Supporting Detail #3

Paragraph # 11 Main Idea: _____ _____	Supporting Detail #1
	Supporting Detail #2
	Supporting Detail #3

G. Hunting for Paragraphs

Find the paragraph that answers each question below. Write the paragraph numbers on the blank lines.

1. Which paragraph explains why Oberth and Von Braun quit working together? _____

2. Which paragraph talks about the first peacetime use for rockets? _____

H. Going Beyond the Main Idea

Go beyond the main ideas in the lesson to answer the following questions.

1. Do you think Von Braun would have stayed in Germany if he had thought Germany would win the war? Explain your reasoning.

   _____

   _____

   _____

2. How does war affect the advancement of science? Give a specific example.

   _____

   _____

   _____

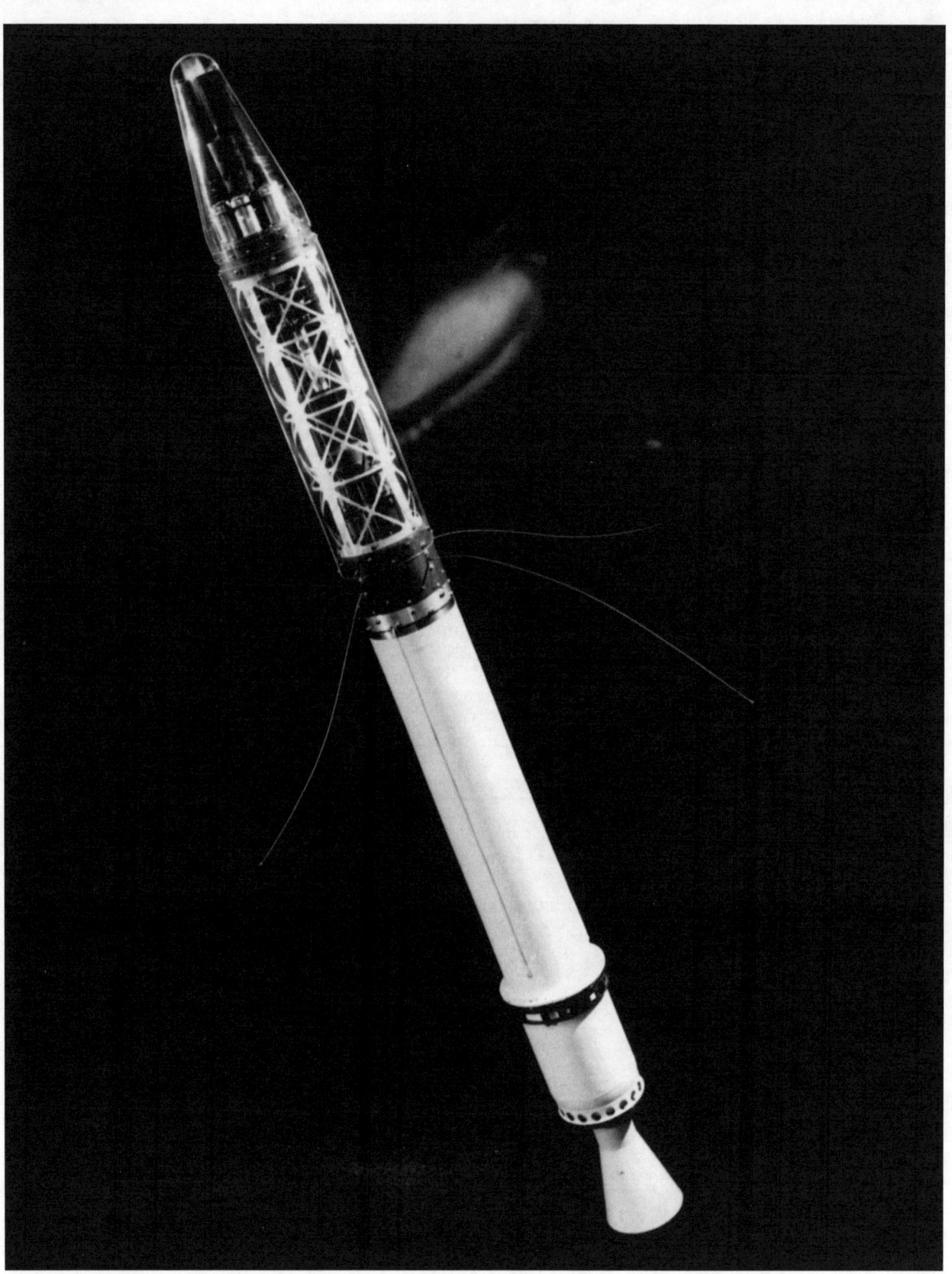

Satellite *Explorer I* was launched into space by the United States on January 31, 1958.

Satellites, The Space Race, and NASA

A. Setting the Stage

Read the first two and last two paragraphs of this lesson. Write four nouns or phrases that appear to be key ideas.

1.	2.	3.	4.

B. Discussing the Background

Reread the last two paragraphs of Lesson 3. As a group, discuss the relationship between Lesson 3 and Lesson 4.

Words to Know
advanced
dominate
frenzy
squabble
orbit
restore
cosmonaut

1 From miles above Earth, *Sputnik I* beeped its electronic signals. The 184-pound *Sputnik* was not an **advanced** satellite. It carried few scientific instruments and took no atmospheric measurements. Its real importance was that it showed the Soviets had led the world into the Space Age.

2 *Sputnik I* should have come as no surprise. In 1955, both the U.S. and the Soviet Union announced they would launch satellites during the International Geophysical Year. This was the eleventh year in the sun's 11-year activity cycle. The window ran from July 1, 1957, to December 31, 1958. The International Council of Scientific Unions called for special ideas and studies. Scientists from 66 countries participated in experiments and programs.

3 In 1955, no one dreamed that the Soviet Union would launch before the United States. After all, the U.S. was the world leader in air power. America's power began with the X-1 that broke the sound barrier in 1947. It continued with the X-15, America's most current rocket plane at that time. U.S. supersonic aircraft **dominated** research and flight. The Soviets' aircraft was not as advanced as America's, but they launched the first rocket.

4 The Soviet Union's national interest in rockets was much older than America's. In fact, the very first theories of rockets, liquid fuels, and space travel had been developed by Russian school teacher Konstantin Tsiolkovsky. He outlined his ideas as early as 1903. Tsiolkovsky was to the Soviet Union what Goddard was to the U.S. and Oberth was to Germany: a father of space and a rocket pioneer.

The Soviets started their space program not long after Oberth published his ideas in Germany. The first Soviet rocket went up shortly after Germany's. Also, about 100 German rocket scientists moved to the Soviet Union after World War II. With their help, the Soviets now appeared to be the leader in space exploration.

5

As if to drive home that new leadership, the Soviet Union launched *Sputnik II* in November of 1957. This satellite weighed more than a thousand pounds and carried a dog!

6

"To explore the phenomenon of the atmosphere and space for peaceful purposes for the benefit of all mankind."
—NASA's objective, stated in the National Aeronautics and Space Act of July 29, 1958

Shock and **frenzy** seized America. What would the Soviets put into space next? A human being? Nuclear weapons? Why had they been able to take America by surprise?

7

The truth was that the U.S. could have launched a satellite a whole year ahead of the Soviet Union! Von Braun and his team were ready with their Redstone rocket. This 68-1/2-foot launch missile was already tested and ready for use.

8

But **squabbles** had arisen among the three branches of the military. The Navy and Air Force both made proposals to develop launch rockets. In September of 1955, the U.S. government gave the go-ahead to the Navy's Viking missile and Project Vanguard.

The Viking's development was full of problems that ate away the time. Meanwhile, the Soviet Union had plodded slowly but steadily ahead, like the turtle in the old story. And, like the turtle, the Soviet Union had reached the goal first. The U.S. immediately set two goals: get a satellite into **orbit,** and organize a space program.

9

The Navy was already scheduled to launch a satellite on December 6, 1957. President Eisenhower turned his attention to organizing a space program. On October 15, Eisenhower appointed a Special Assistant for Science and Technology. This appointment came just eleven days after *Sputnik I* made it into orbit.

10

On December 6, 1957, Americans gathered to watch a Viking rocket launch the United States' first satellite. The hope of **restored** American confidence and spirit was squashed in embarrassment. The rocket exploded on its launch pad, and the satellite rolled across the ground.

11

Fortunately, Von Braun's Project Orbiter was ready. On January 31, 1958, Von Braun's Jupiter C launched American satellite *Explorer I* into orbit. The Navy readied a second Viking rocket. On March 17, 1958, satellite *Vanguard I* joined *Explorer I.*

12

In July, Congress passed the National Aeronautics and Space Act of 1958. By October, the National Aeronautics and Space Administration (NASA), was up and running. During the next three years, NASA led the U.S. in its recovery from "Sputnik shock." Project Score broadcast the first voiced message from space. *Vanguard II* sent back the first weather data. *Tiros I* took the first detailed weather pictures. *Transit IB,* collected the first navigation guides for pilots and sailors.

13

On December 4, 1959, NASA safely launched and returned a monkey named "Sam" in space. The first astronauts were training, and NASA built the first space capsule. By 1961, it seemed as if the U.S. was even in the space race with the Soviets.

14

Then the Soviets shot ahead again. On April 12, 1961, they launched **cosmonaut** Yuri Gagarin into space.

15

C. Finding the Main Idea

Highlight or circle the main idea in each paragraph. Remember that the main idea of a paragraph is often the first or last sentence in that paragraph. Other times, the main idea has to be pieced together from more than one sentence.

D. Making a Timeline

Place three important dates from this lesson on the timeline. Also place one important date from Lesson 3. Write two or three words to identify the importance of each date.

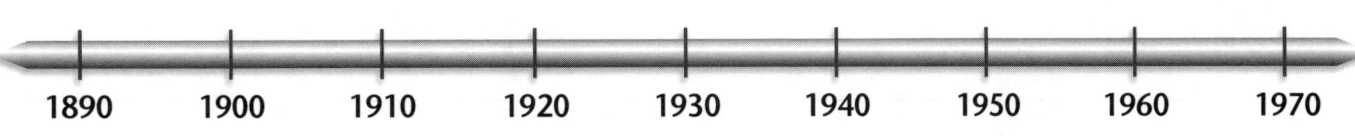

| 1890 | 1900 | 1910 | 1920 | 1930 | 1940 | 1950 | 1960 | 1970 |

E. Using Context Clues

Find the following vocabulary words in the lesson. The words and sentences around each word give a clue to its meaning. Use these clues to write a meaning for each word. For each vocabulary word, describe one clue that you used.

1. dominated

Meaning: _____

_____

_____

Clue: _____

_____

2. frenzy

Meaning: _____

_____

_____

Clue: _____

_____

F. Reading for Details

Reread the paragraph identified below. Then, complete the main idea/detail web.

Paragraph # 13 Main Idea: _____ _____ _____	Supporting Detail #1
	Supporting Detail #2
	Supporting Detail #3

G. Charting the Race

Complete this chart by filling in the correct dates in the squares. Use information through April 1961. Put an X in the box if the event had not happened yet.

	United States	Soviet Union
Sent a person into space		
Sent an animal into space		
Sent a satellite into space		

Identifying Cause and Effect

Mercury capsule
Height: 6⁷/₈ feet
Crew: 1

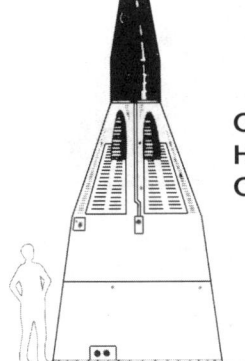

Gemini capsule
Height: 19 feet
Crew: 2

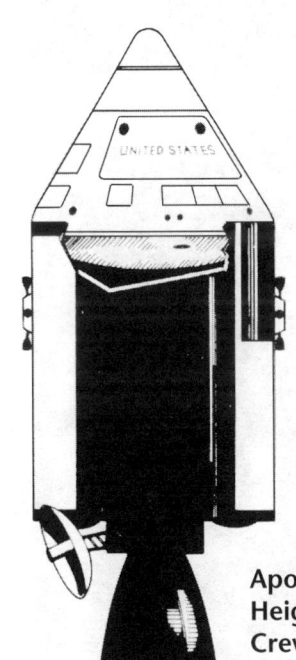

Apollo Command Module
Height: 35¹/₃ feet
Crew: 3

Once you have learned to identify main ideas and details, you can use this skill to recognize causes and their effects. Recognizing causes and effects helps you understand and connect information in a reading selection.

A *cause* is an event, action, happening, or kind of behavior that directly leads to another event, action, or kind of behavior.

An *effect* is an event, action, happening, or behavior that is the outcome or result of an event, action, happening, or behavior.

How to recognize implied cause and effect:

1. Read each paragraph.

2. Ask yourself "What happened?" This helps you recognize effect. (Often, the main idea of a paragraph is an effect.)

3. Then, ask yourself "Why did this happen?" This helps you recognize cause. (Often, details will explain cause.)

4. Sometimes, signal words will serve as clues that a cause-and-effect relationship exists. Some common signal words are *because, therefore, so, as a result,* or *since.* Often cause-and-effect relationships do not use signal words. But, you can insert one of these words to check the connection between a main idea and details. If the connecting word makes sense, you probably have a cause-and-effect relationship.

Mercury astronaut John Glenn and his capsule, *Friendship 7,* before flight on February 20, 1962.

Project Mercury

A. Setting the Stage

Read the first two and last two paragraphs of this lesson. Write four nouns or phrases that appear to be key ideas.

1.	2.	3.	4.

B. Discussing the Background

Reread the last two paragraphs of Lesson 4. As a group, discuss the relationship between Lesson 4 and Lesson 5.

Words to Know

respond

superior

puncture

meteoroid

reentry

1 Just as they had been by *Sputnik I,* Americans were upset by another Soviet first: a cosmonaut circling the Earth. NASA's goal was to get a person in orbit. The whole country was reading about Project Mercury, NASA's program to reach that goal. The U.S. people wondered how the Soviet Union had beaten them to the goal.

2 NASA was careful—too careful, many critics said. But NASA did not want to risk a mishap on the first launch of a person into space. Safety was very important to NASA. So NASA tested and retested machinery and astronauts. Both were very near launch-ready when Gagarin went up. So the U.S. was able to **respond** rapidly to the cosmonaut in orbit.

3 On May 5, 1961—just 23 days after Gagarin's flight—U.S. astronaut Alan Shepard rocketed into space. As with its satellite launch, America's launch of a person into space was second, but **superior** to the Soviet Union's. Commander Shepard controlled his spacecraft, *Freedom 7,* during his 15-minute flight. Gagarin was simply a passenger during his flight. The general public did not know of his flight until success was certain. Television cameras followed Shepard's flight, for all the world to see.

4 Getting Project Mercury off the ground and Shepard into space was no easy feat. The task required NASA to develop a new launch rocket, a new spacecraft, and to find and train a new kind of pilot.

Fortunately, NASA did not have to start from scratch. It could draw from past test programs of the Army, Navy, and Air Force. NASA shared data and designs with companies such as General Electric and McDonnell Aircraft, and universities such as the California Institute of Technology.   5

To design a new space vehicle, NASA scientists began with data from other launches. They looked at both the supersonic X-15 rocket plane and the V-2 upper atmosphere launches of the early 1950s. To design a new launch rocket, NASA began with Von Braun's Redstone from Project Orbiter, and the Air Force's Atlas rockets. As for its new pilots, NASA turned to the Navy and Air Force.   6

"I believe that this nation should commit itself to achieving the goal before this decade is out, of landing a man on the moon and returning him safely to Earth."
—President John F. Kennedy, before a joint session of Congress, May 25, 1961

Of course, some of the available data was not useful. Project Mercury was truly going "where no human had gone before." NASA had to find answers to a thousand new questions. Astronauts had to be trained and a flying machine had to be built to handle imagined problems and conditions.   7

The astronauts had to learn to withstand the ear-shattering noise and pounding vibration of blast-off. They had to know how to handle the pressure of traveling through Earth's atmosphere against gravity's pull. The sudden weightlessness of space required more special adaptations.   8

The machine had to be strong enough to survive blast-off. The outer material needed to be strong enough to resist **puncture** by whizzing **meteoroids** in space. After surviving all of this, the machine had to be able to get through **reentry** without melting.   9

The flight of Alan Shepard and Mercury capsule *Freedom 7* was flawless. Shepard suffered no ill effects going up or coming down. *Freedom 7* worked perfectly—all 120 controls, 55 electrical switches, 30 fuses, 35 mechanical levers, and 7 miles of wiring!   10

Freedom 7's success had several positive results. Shepard became an American hero and *astronaut* became a household word. Twenty days later, on May 25, 1961, President Kennedy spoke to Congress. He wanted the United States to commit to putting a person on the moon before 1970.   11

NASA, however, still needed to send a person into orbit. The Soviet Union launched a second cosmonaut on August 6, 1961. Gherman Titov orbited the earth 17 times. More loudly than ever, Americans called for a speed-up in the Project's launches.

12

NASA introduced Atlas, a more powerful launch rocket. This change meant more tests to prove the Mercury capsule could handle higher speeds and temperatures. Finally, on February 20, 1962, John Glenn became the first American to orbit Earth. Glenn circled Earth three times at five miles per second. He also piloted the capsule when the automatic controls failed. Gagarin had not been equipped to pilot his craft.

13

NASA launched three more Mercury missions. Each was longer and more complicated than the one before. Astronaut Gordon Cooper's 34-hour, 22-orbit flight on May 15, 1963, closed the door on a successful Project Mercury. NASA had safely flown a person into space and back.

14

NASA began the second phase of the journey to the moon. Project Mercury had put an American in orbit. Project Gemini would accomplish a second wonder: the first American space walk.

15

C. Finding the Main Idea

Highlight or circle the main idea in each paragraph. Remember that the main idea of a paragraph is often the first or last sentence in that paragraph. Other times, the main idea has to be pieced together from more than one sentence.

D. Making a Timeline

Place three important dates from this lesson on the timeline. Also place one important date from Lesson 4. Write two or three words to identify the importance of each date.

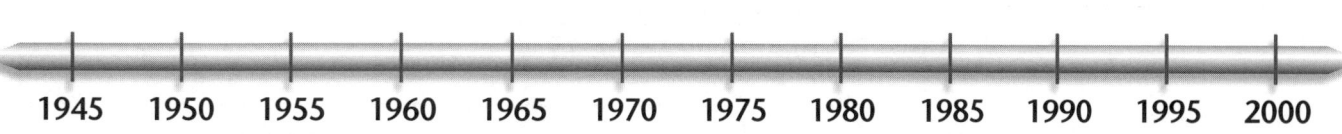

1945	1950	1955	1960	1965	1970	1975	1980	1985	1990	1995	2000

E. Using Context Clues

Find the following four vocabulary words in the lesson. The words and sentences around each word give a clue to its meaning. Use these clues to write a meaning for each word. For each vocabulary word, describe one clue that you used.

1. respond

Meaning: _____

_____

_____

Clue: _____

_____

2. superior

Meaning: _____

_____

_____

Clue: _____

_____

3. puncture

Meaning: _____

_____

_____

Clue: _____

_____

4. meteoroids

Meaning: _____

_____

_____

Clue: _____

_____

F. Reading for Details

Reread the two paragraphs identified below. Then, complete the main idea/detail webs.

Paragraph # 6 Main Idea: _____ _____	Supporting Detail #1
	Supporting Detail #2
	Supporting Detail #3

Paragraph # 10 Main Idea: _____ _____	Supporting Detail #1
	Supporting Detail #2
	Supporting Detail #3

Project Mercury

G. Identifying Cause and Effect

Finish the Cause and Effect structures below. Begin by rereading the paragraphs. Keep in mind that an *effect* is something that happened. A *cause* tells why something happened. To check that you have identified a cause-and-effect relationship, create a cause-and-effect sentence. Read the effect, then the word *because,* then the cause. If this sentence makes sense, you have correctly identified a cause-and-effect relationship.

1. Paragraph 1

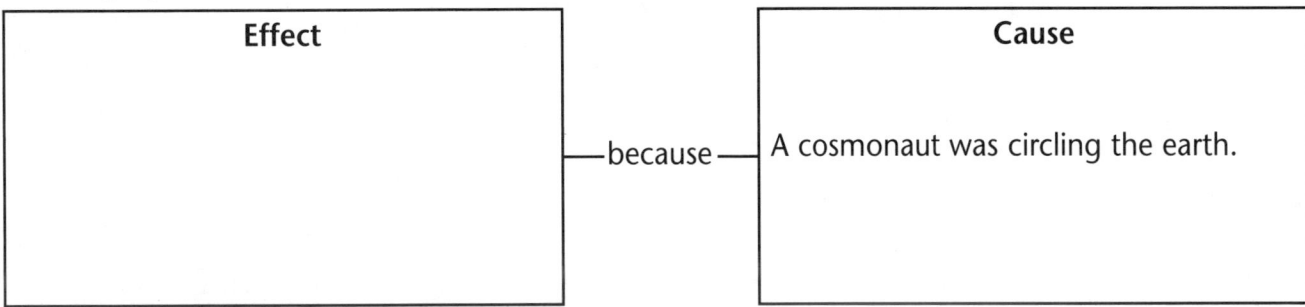

Effect		Cause
	—because—	A cosmonaut was circling the earth.

2. Paragraph 7

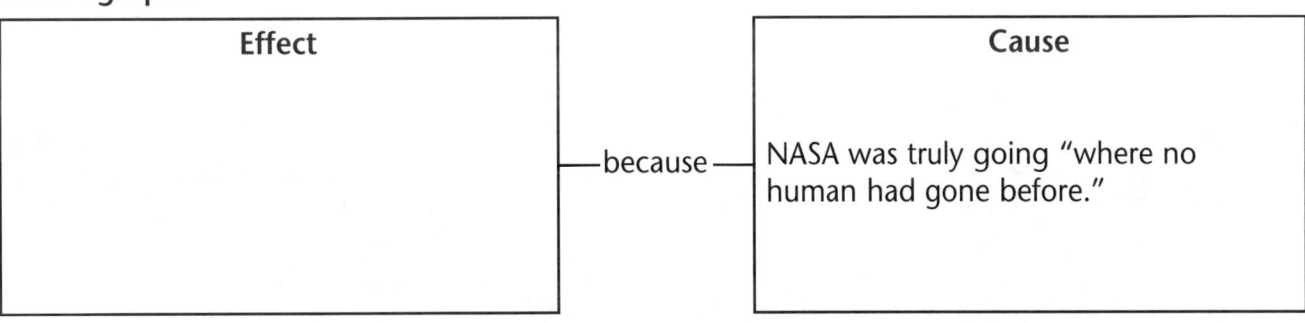

Effect		Cause
	—because—	NASA was truly going "where no human had gone before."

3. Paragraph 12

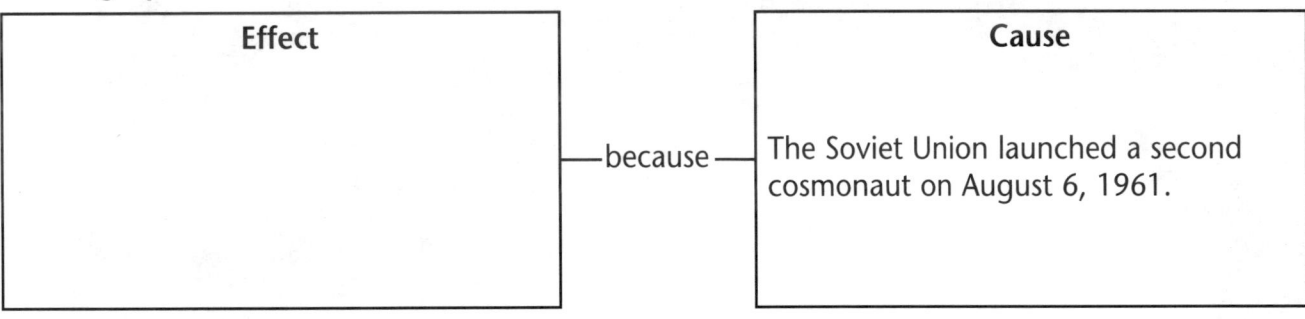

Effect		Cause
	—because—	The Soviet Union launched a second cosmonaut on August 6, 1961.

Edward White makes the first spacewalk of America's space program. He is connected to the Gemini spacecraft by a gold-plated cord.

A. Setting the Stage

Read the first two and last two paragraphs of this lesson. Write four nouns or phrases that appear to be key ideas.			
1.	2.	3.	4.

B. Discussing the Background

Reread the last two paragraphs of Lesson 5. As a group, discuss the relationship between Lesson 5 and Lesson 6.

Words to Know

rendezvous

extravehicular

suborbital

maneuver

propel

disengage

recap

fatality

1 The two years between the completion of Project Mercury and the beginning of Project Gemini were years of growth. Every part of America's space program grew bigger.

2 One goal for Mercury had become three for Gemini: complete longer flights, **rendezvous** and dock two spacecraft in orbit, and perform **extravehicular** activities. To meet these goals, the number of missions grew: NASA scheduled ten flights for Project Gemini.

3 Gemini's missions called for two astronauts instead of Mercury's one. NASA's astronaut corps grew. Two groups of new astronauts reported for space training, one group of nine signing on in 1962, another of 14 in 1963.

4 Two-person crews needed more room. The capsule had to grow. Gemini ended up about three times bigger than Mercury. The craft retained its special bell shape. Much of the equipment that took up room inside Mercury was moved to the outside of Gemini. In every sense of the word, the Gemini vehicle was a spacecraft. The term *capsule* disappeared from space vocabulary.

5 Because the spacecraft grew, so did the launching rocket. The Redstone and Atlas used in Project Mercury were about 6 feet in diameter. They could launch payloads weighing up to a ton. The Gemini craft was three times that heavy and 7-1/2 feet in diameter. A new rocket, the Titan II, was added to the team.

NASA launched Project Gemini with two test flights. Neither flight had a crew of astronauts. The first lifted off on April 8, 1964. The three-orbit flight demonstrated that the Gemini-Titan II combination would work. The second, on January 19, 1965, was a **suborbital** flight to prove that the spacecraft's heat shield could protect it during reentry.

6

Then, almost immediately, NASA paired its machinery with its two-astronaut crews and Project Gemini really took off. In 1965, Gemini astronauts flew three missions and met two of the Project's goals.

7

"The [Gemini spacecraft] was needed to answer the three questions of rendezvous, long duration, and extravehicular operation."
—Michael Collins, *Liftoff*

Astronauts Gus Grissom and John Young blasted off on March 23 to test Gemini's systems. In particular, they were to try the space program's first **maneuvers** in orbit. By firing a series of small rockets, they moved the craft in and out of orbit, and also changed its position. This ability to move around was necessary for future rendezvous and docking.

8

Less than three months after Grissom's and Young's flight, astronauts Ed White and James McDivitt took to the skies to meet one of Project Gemini's goals: extravehicular activity. On June 3, White took the first American space walk out of, then around, over, and under his ship. Only a slender, gold-plated cord ran between White and the spacecraft, piloted by McDivitt. For about 20 minutes, White tested a hand-maneuvering unit. This gun **propelled** him by squirting out gas jets. Then its fuel ran out. White was enjoying his walk so much that he did not want to stop. Finally, in "the saddest moment of my life," White rejoined McDivitt inside *Gemini 4*.

9

In December of 1965, astronauts carried out the first space rendezvous. Astronauts Frank Borman and James Lovell took off in *Gemini 7* on December 4. On December 15, astronaut Wally Schirra piloted *Gemini 6* to meet *Gemini 7* in orbit, 182 miles above Earth. There, speeding at 17,500 miles per hour and only one foot apart, the two crafts moved over, under, and around each other. Schirra and his craft then returned to Earth. Borman and his copilot, James Lovell, remained in orbit for a two-week space stay. Borman and Lovell met another of the Project's goals, as well as set a record. Their stay was to be the longest in all three Projects.

10

Project Gemini

On March 16, 1966, astronaut Neil Armstrong tested the docking of *Gemini 8*. Several hours after starting to orbit, Armstrong guided his craft into the docking collar of an Agena rocket. He locked the two craft together and orbited smoothly for about 40 minutes. Suddenly the vehicles began to jerk. Armstrong **disengaged** *Gemini 8* and made an emergency landing. The jerking problem turned out to be minor and the mission was a success. Armstrong had been able to dock the Gemini almost as easily as parking a car.

11

NASA wrapped up Project Gemini with four more successful missions. The missions **recapped** procedures learned on the previous six. For the most part, Gemini's equipment worked well. The crews were able to fix the problems that came up. On November 15, 1966, NASA wound down Project Gemini and began gearing up for Project Apollo.

12

Project Apollo was grounded before it even got started. On January 27, 1967, an electrical fire swept the command module of *Apollo 1*. Trapped inside were three astronauts: Mercury and Gemini veteran Gus Grissom, spacewalker Ed White, and rookie Roger Chaffee. They were the first **fatalities** of America's space program.

13

C. Finding the Main Idea

Highlight or circle the main idea in each paragraph. Remember that the main idea of a paragraph is often the first or last sentence in that paragraph. Other times, the main idea has to be pieced together from more than one sentence.

D. Making a Timeline

Place three important dates from this lesson on the timeline. Also place one important date from Lesson 5. Write two or three words to identify the importance of each date.

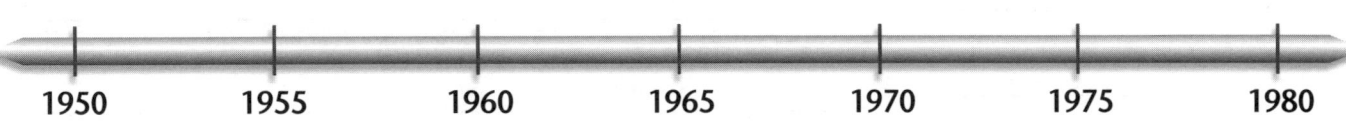

| 1950 | 1955 | 1960 | 1965 | 1970 | 1975 | 1980 |

E. Using Context Clues

Find the following four vocabulary words in the lesson. The words and sentences around each word give a clue to its meaning. Use these clues to write a meaning for each word. For each vocabulary word, describe one clue that you used.

1. extravehicular

Meaning: _____

_____

_____

Clue: _____

_____

2. maneuvers

Meaning: _____

_____

_____

Clue: _____

_____

3. disengaged

Meaning: _____

_____

_____

Clue: _____

_____

4. fatalities

Meaning: _____

_____

_____

Clue: _____

_____

F. Reading for Details

Reread the two paragraphs identified below. Then, complete the main idea/detail webs.

Paragraph # 8

Main Idea: _____

_____

_____

Supporting Detail #1

Supporting Detail #2

Supporting Detail #3

Paragraph # 11

Main Idea: _____

_____

_____

Supporting Detail #1

Supporting Detail #2

Supporting Detail #3

G. Identifying Cause and Effect

Finish the Cause and Effect structures below. Begin by rereading the paragraphs. Keep in mind that an *effect* is something that happened. A *cause* tells why something happened. To check that you have identified a cause-and-effect relationship, create a cause-and-effect sentence. Read the effect, then the word *because,* then the cause. If this sentence makes sense, you have correctly identified a cause-and-effect relationship.

1. Paragraph 4

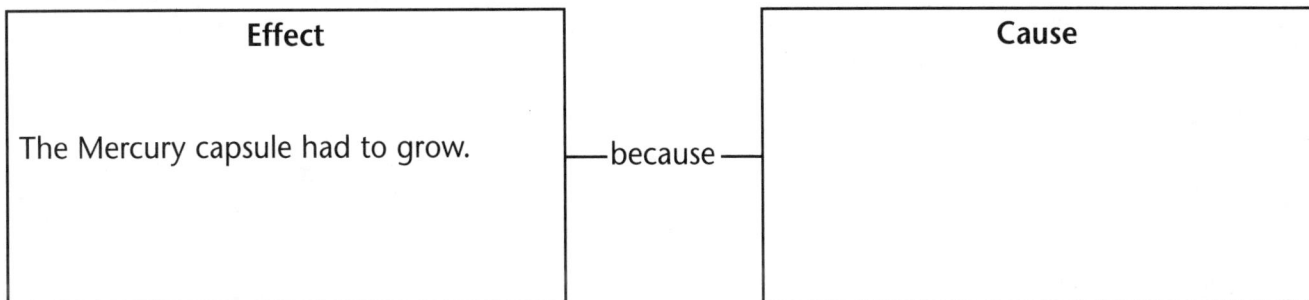

Effect		**Cause**
The Mercury capsule had to grow.	—because—	

2. Paragraph 10

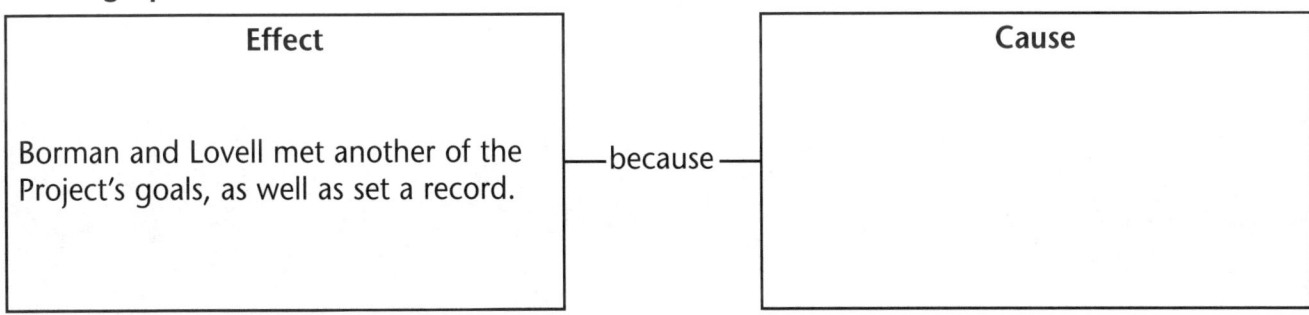

Effect		**Cause**
Borman and Lovell met another of the Project's goals, as well as set a record.	—because—	

3. Paragraph 14

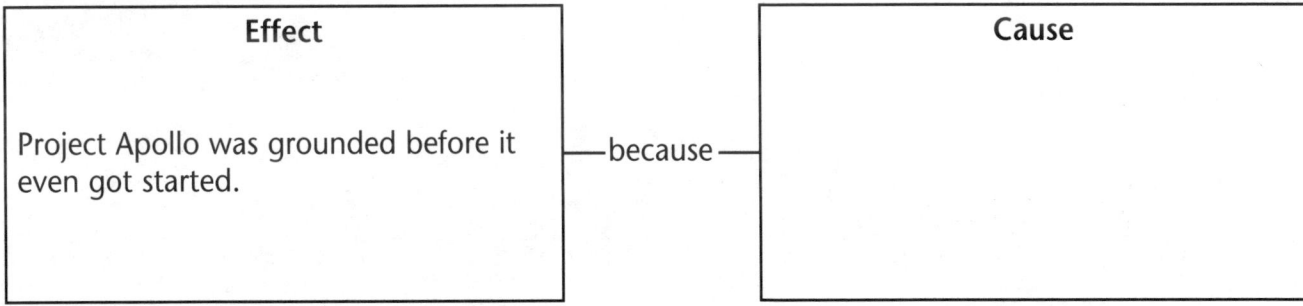

Effect		**Cause**
Project Apollo was grounded before it even got started.	—because—	

Astronaut Edwin "Buzz" Aldrin releases a solar wind experiment from *Apollo 11.*

Project Apollo

A. Setting the Stage

Read the first two and last two paragraphs of this lesson. Write four nouns or phrases that appear to be key ideas.

1.	2.	3.	4.

B. Discussing the Background

Reread the last two paragraphs of Lesson 6. As a group, discuss the relationship between Lesson 6 and Lesson 7.

Words to Know

lunar

site

bleak

malfunction

splashdown

ingenuity

awe

humdrum

1 NASA readied Project Apollo for early 1967 by sending **lunar** probes to the moon. While the Gemini teams spacewalked and docked, the probes looked for a landing **site** for the Project Apollo spacecraft. The *Ranger 7* probe (1964) sent back more than 4,000 photos. Surveyor probes followed the Rangers. Surveyors made test landings on the moon to prove the surface could support a spacecraft. *Lunar Orbiter 1* was launched on August 10, 1966. It sent back more than 400 photos.

2 All plans were put on hold after that **bleak** day in January of 1967. The cause of *Apollo 1's* fire was never determined. However, the investigators found some problems that probably added to the disaster. Before Project Apollo continued, more than a thousand changes were made to the spacecraft.

3 During the 22 months of redesign, words of Gus Grissom rang silently in the air. "We are in a risky business, and we hope if anything happens to us it will not delay the program." Finally, on October 11, 1968, a huge new Saturn rocket blasted a renewed Apollo spacecraft into orbit. NASA was back on its way to the moon.

4 The Apollo spacecraft had two parts. The larger section, the command module, housed the three astronauts, equipment, and supplies. The lunar module was a much smaller section, built to carry two astronauts between the moon's surface and the orbiting command module.

Early missions of Project Apollo checked out the new spacecraft. Within Earth's orbit, the crews of *Apollo 7* and *Apollo 8* tested the command module. Then, on December 21, 1968, *Apollo 8* was launched. The crew was headed toward the moon on a scouting trip. Two-hundred forty thousand miles out, 10 orbits around the moon, 240,000 miles back: the command and service module had no trouble during the six-day mission.

5

By March of 1969, the lunar modules were ready for testing. Nicknamed "Spider," the little sections were mated with command sections. *Apollo 9* and *Apollo 10* were ready to go. Like the command module, the lunar module tests in Earth and lunar orbits were problem-free.

6

On July 16, 1969, American astronauts Neil Armstrong, Edwin "Buzz" Aldrin, and Michael Collins blasted off in *Apollo 11*. Four days later, they were orbiting the moon. Armstrong and Aldrin left *Apollo 11's* command module *Columbia* and entered lunar module *Eagle*. Rockets fired *Eagle* towards the moon. Nine minutes later Armstrong's voice was heard on Earth: "Houston, **Tranquility** Base here. The *Eagle* has landed."

7

"That's one small step for [a] man, one giant leap for mankind."
—Neil Armstrong, from the moon, July 20, 1969

On July 20, 1969, American astronaut Neil Armstrong became the first person to set foot on the moon. Armstrong walked in fine, powdery moon dust for about 20 minutes. Then, fellow astronaut Buzz Aldrin joined him. For two hours, the men collected rock and soil samples. They set up scientific instruments. They left their footprints and a small plaque that read:

8

Here men from the planet Earth first set foot upon the moon, July 1969 A.D. We came in peace for all mankind.

More than 500 million people on Earth watched Armstrong and Aldrin make a giant leap for America and for the whole human race.

Six more Apollo crews set out for the moon. Only five made it. The crew of *Apollo 13* nearly died in space when the fuel cells **malfunctioned**. With **ingenuity** and courage, the three men powered down the command module and closed it off. They crawled into the tiny lunar module, where they stayed for 90 cold, cramped hours. The space was designed for only two people and for only 45 hours. The lunar module also towed the command module back to Earth for reentry and **splashdown!**

9

Apollo 15's crew used a new, advanced piece of space machinery, the *Lunar Rover*. The *Rover* was a 500-pound, fold-down vehicle with wire-mesh wheels. It looked a lot like a dune buggy, and made "cruising the moon" much easier. There is much less gravity on the moon than on Earth, and bounding around could be tiring. Also, the crew's suits were bulky and their movements were somewhat clumsy. The *Lunar Rover* raced over the moon at speeds up to 11 miles per hour.

10

11

By the end of Project Apollo, 12 people had walked on the moon. All 12 were American astronauts. They gathered more than 800 pounds of moon rocks and soil, took thousands of pictures, and conducted hundreds of scientific experiments. But public admiration and **awe** had died. What was one of the greatest accomplishments of human exploration had become **humdrum**. By the close of Project Apollo, many people were bored with moon landings. Some even called local TV stations to complain that coverage of the landings was interrupting their favorite programs!

Was the heyday of space exploration already over? Would future history books tell of the "golden age of space flight: October 1957 to December 1972"? Not if NASA had anything to say about it! In May of 1973, five months after the final Apollo flight, NASA launched *Skylab*. America's first experimental space station started a new chapter in space history!

12

C. Finding the Main Idea

Highlight or circle the main idea in each paragraph. Remember that the main idea of a paragraph is often the first or last sentence in that paragraph. Other times, the main idea has to be pieced together from more than one sentence.

D. Making a Timeline

Place three important dates from this lesson on the timeline. Also place one important date from Lesson 6. Write two or three words to identify the importance of each date.

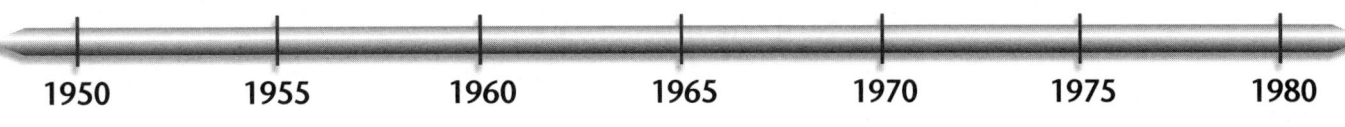

| 1950 | 1955 | 1960 | 1965 | 1970 | 1975 | 1980 |

E. Using Context Clues

Find the following four vocabulary words in the lesson. The words and sentences around each word give a clue to its meaning. Use these clues to write a meaning for each word. For each vocabulary word, describe one clue that you used.

1. lunar

Meaning: _____

_____

_____

Clue: _____

_____

2. malfunctioned

Meaning: _____

_____

_____

Clue: _____

_____

3. ingenuity

Meaning: _____

_____

_____

Clue: _____

_____

4. humdrum

Meaning: _____

_____

_____

Clue: _____

_____

F. Reading for Details

Reread the paragraphs identified below. Then, complete the main idea/detail webs.

Paragraphs # 7–8

Main Idea: _____

_____

_____

Supporting Detail #1

Supporting Detail #2

Supporting Detail #3

Paragraph # 9

Main Idea: _____

_____

_____

Supporting Detail #1

Supporting Detail #2

Supporting Detail #3

Project Apollo

G. Identifying Cause and Effect

Finish the Cause and Effect structures below. Begin by rereading the paragraphs. Keep in mind that an *effect* is something that happened. A *cause* tells why something happened. To check that you have identified a cause-and-effect relationship, create a cause-and-effect sentence. Read the effect, then the word *because,* then the cause. If this sentence makes sense, you have correctly identified a cause-and-effect relationship.

1. Paragraph 1

Effect		Cause
	—because—	NASA needed to prove the surface of the moon could support a spacecraft.

2. Paragraph 9

Effect		Cause
The three men powered down the command module and closed it off.	—because—	

H. Looking at the Big Picture

On the left, number the spacecrafts in order of appearance in the NASA program, starting with the earliest. Then, on the right, rewrite the list of crafts in the correct order.

_____ Ranger	1. _____
_____ Apollo	2. _____
_____ Mercury	3. _____
_____ Lunar Rover	4. _____
_____ Gemini	5. _____
_____ Skylab	6. _____
_____ Lunar Orbitor	7. _____
_____ Surveyor	8. _____

America's first space station, *Skylab,* was the largest spacecraft put into orbit when launched on May 14, 1973. Above is *Skylab 4.*

Skylab and Rocket Planes

A. Setting the Stage

Read the first two and last two paragraphs of this lesson. Write four nouns or phrases that appear to be key ideas.

1.	2.	3.	4.

B. Discussing the Background

Reread the last two paragraphs of Lesson 7. As a group, discuss the relationship between Lesson 7 and Lesson 8.

Words to Know

achieve

convert

outfit

extensive

mass

simulated

delta wings

1 Ideas for a space station had always gone hand-in-hand with dreams of flight and space travel. Such stations seemed a logical first step in space exploration. From them, scientists could look back and study Earth. Space expeditions could blast off much more easily from a space station than from Earth. In fact, in 1958, when NASA started, plans called for a space station to be built along with the first rocket.

2 The space race and President Kennedy's challenge changed NASA's plans. Any ideas that did not directly contribute to a moon landing were put on hold.

3 Armstrong and Aldrin **achieved** the President's goal. As soon as *Apollo 11* had completed its successful moon landing, NASA took the space station plans from the shelf, blew off the dust, and went to work on *Skylab*.

4 *Skylab* was a **converted** Saturn V, one of the huge rockets that launched Apollo spacecraft. The rocket's first two stages stayed the same, as fuel tanks to launch *Skylab* into orbit. The rocket's third stage, a huge cylinder 48 feet long and 22 feet in diameter, became *Skylab*. It was **outfitted** with a special shield that protected the outside from damage and kept the inside cool. *Skylab* had four small solar panels on one end, and two large solar wings. These tools collected the sun's energy and turned it into electrical power. Inside, *Skylab* had two floors: living quarters and a space laboratory.

On May 14, 1973, just five months after the last Apollo moon landing, *Skylab* was launched into orbit. Although it was damaged during launch, the first crew was able to make repairs. *Skylab* hosted crews for three long space stays that lasted a total of 171 days.

5

During this time, NASA carried out **extensive** medical testing of the astronauts. The tests showed that long-term space stays did not harm their long-term health. Muscles, including the heart, did begin to weaken. Bones lost some **mass** and strength. But NASA specialists found that the body gradually resumed its pre-flight state upon return to Earth. They concluded that long space stays might require **simulated** gravity.

6

Extensive medical data was only a fraction of the information that *Skylab* collected. Its cameras returned more than 200,000 photographs of the sun's activities, from solar winds to flares. Also captured on film were previously unseen ultraviolet and X-ray wavelengths. They had always been blocked from Earth's cameras by the atmosphere. Photographs of Earth's oceans and deserts provided scientists with much valuable information.

7

> "The real payoff—the reason for the whole *Skylab* project—is the data we've obtained."
> —William Schneider, *Skylab* director, 1968–1974

NASA planned a fourth mission to *Skylab*. But the sun's activity affected the atmosphere at the space station's altitude. This change caused the orbit of the lab to decay. On July 11, 1979, *Skylab* fell from orbit and was destroyed when it reentered Earth's atmosphere. Fortunately, the tremendous amount of data it had collected had already passed safely back to Earth. Scientists and engineers were applying it to another one of NASA's plans: a reusable rocket plane.

8

Like the dreams of a space station, dreams of a rocket plane that could travel into space had been around for a long time. Tsiolkovsky, Goddard, and Oberth had all put forth plans for such a vehicle. The U.S. began work on rocket planes in the middle 1940s. Charles Yeager broke the sound barrier in a rocket plane, the X-1.

9

Through the 1950s, a number of rocket planes were developed from the X-1. By 1962, when John Glenn orbited the earth, the X-1 had given birth to the X-15. The X-15 was a 30,000-pound wonder that flew at six times the speed of sound and could reach altitudes of 60 miles. (The edge of space is about 62 miles above the surface of the earth.)

10

During the 1960s, public attention was fastened on the doings of NASA's Space Division. But its Aeronautics Division continued to look for the perfect rocket plane. By the time Armstrong walked on the moon, the X-20 Dyna Soar was in development. The development team included NASA, the Air Force, and the Boeing Company. The X-20 had unusual wings: huge triangular shapes running almost the length of the plane's body. These wings were called **delta wings.** Tests showed that these wings gave a rocket plane better handling qualities. They also made the approach for landing, as well as the landing itself, safer.

11

The X-20 Dyna Soar never flew. Instead, NASA combined qualities of the X-15 and the X-20 Dyna Soar. This combination resulted in a reusable rocket plane. Data from *Skylab* was used to refine the new bird. All told, the craft cost $9 billion and came off the assembly line more than two years behind schedule. But it excited America as nothing in the air had since *Apollo 11.* By 1981, everybody knew about NASA's reusable rocket plane, the *space shuttle.*

12

C. Finding the Main Idea
Highlight or circle the main idea in each paragraph. Remember that the main idea of a paragraph is often the first or last sentence in that paragraph. Other times, the main idea has to be pieced together from more than one sentence.

D. Making a Timeline
Place three important dates from this lesson on the timeline. Also place one important date from Lesson 7. Write two or three words to identify the importance of each date.

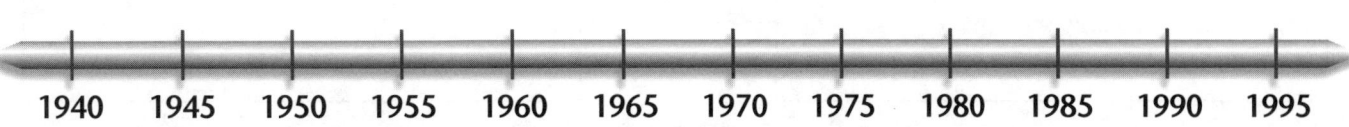

1940   1945   1950   1955   1960   1965   1970   1975   1980   1985   1990   1995

E. Using Context Clues

Find the following vocabulary words in the lesson. The words and sentences around each word give a clue to its meaning. Use these clues to write a meaning for each word. For each vocabulary word, describe one clue that you used.

1. outfitted

Meaning: _____

_____

_____

Clue: _____

_____

2. extensive

Meaning: _____

_____

_____

Clue: _____

_____

F. Reading for Details

Reread the paragraph identified below. Then, complete the main idea/detail web.

Paragraph # 5 Main Idea: _____ _____ _____	
	Supporting Detail #1
	Supporting Detail #2
	Supporting Detail #3

G. Identifying Cause and Effect

Finish the Cause and Effect structures below. Begin by rereading the paragraphs. Keep in mind that an *effect* is something that happened. A *cause* tells why something happened. To check that you have identified a cause-and-effect relationship, create a cause-and-effect sentence. Read the effect, then the word *because,* then the cause. If this sentence makes sense, you have correctly identified a cause-and-effect relationship.

1. Paragraph 2

Effect Ideas not having to do with the moon landing were put on hold.	—because—	Cause

2. Paragraph 8

Effect *Skylab* fell from the sky.	—because—	Cause

Outlining the Text

Space shuttle *Columbia* heads for space on April 12, 1981. *Columbia's* launch brought America into a new space age.

An outline is a way to present information visually. By outlining material you have read, you can accomplish the following tasks:

1. Identify main ideas and details

2. Organize information

3. Clarify information

4. Summarize information

5. See relationships and make connections

An outline can vary in form from a formal sentence outline to a simple phrase or word outline. Because an outline is a tool, it can take the form most suited to the user's purpose.

In the first eight lessons, you completed main idea/detail webs. Without realizing it, you were outlining paragraphs in each article. Now, you will use the webs to outline an entire article.

Space shuttle *Columbia's* main wheels touch the ground to successfully complete its sixth mission. The shuttle lands horizontally, like an airplane.

The Space Shuttle Flies

A. Setting the Stage

Read the first two and last two paragraphs of this lesson. Write four nouns or phrases that appear to be key ideas.

1.	2.	3.	4.

B. Discussing the Background

Reread the last two paragraphs of Lesson 8. As a group, discuss the relationship between Lesson 8 and Lesson 9.

Words to Know

feature

function

engulf

complex

fragile

wing-span

billow

1 The shining, white craft waiting for launch on April 12, 1981, was unusual. It combined **features** and **functions** of almost every air or spacecraft ever flown. At the same time, it was different from any air or spacecraft ever flown.

2 The shuttle took off like a rocket, orbited like a satellite, descended like a glider, and landed like an airplane. What made it new and different was its reusability. After a two or three-week check-over, the shuttle could fly into space again, and again, and again.

3 Of all its features, its reusability was the most outstanding. Space capsules, spacecraft, rockets, and satellites were always destroyed coming down through Earth's atmosphere. The shuttle was the first craft that was able to go back up again. Reusability was the feature that made it possible for NASA to gain approval and funding during a time when America faced many demands for dollars.

4 NASA developed the shuttle during the late 1960s and early 1970s. America was fighting an expensive and unpopular war in Vietnam. The civil rights movement was **engulfing** the country from coast to coast, border to border. Poverty in American cities was growing and becoming more and more a concern, as was the use of drugs.

5 Solving these problems would take dollars. NASA suddenly found itself facing money problems. America no longer was willing to spend billions on a throw-away space program. NASA worked to present a more practical, economical space program.

NASA's solution was the shuttle. It was reasonably priced, as spacecraft go: a yearly development budget between $5 and $6 billion. It was practical: it could go to space again and again. And it was imaginative: NASA saw it as the first step to a working space station. In time, bases could be set up on the moon and Mars.

6

President Nixon approved NASA's shuttle design on January 5, 1972. But Presidential approval one year did not necessarily mean congressional approval every year. Every year for the next nine years, NASA and Congress warred over funds. NASA very seldom received the annual amount that Nixon had okayed.

7

". . . with the first flight of the space shuttle, America began a new era . . . . Our astronauts could . . . conduct experiments and observations of Earth and the universe unparalleled since the beginning of the space age."
—James M. Beggs, NASA Administrator, 1981–1985

Often, this lack of funding led to delays in the shuttle's development. But just as often, delays also happened because the shuttle was such a **complex** machine. A small change in one part of the shuttle meant changes in other parts and systems.

8

By April 12, 1981, the fights had been fought and the delays were over. Space shuttle *Columbia* hung strapped to its 154-foot fuel tank, ready to launch. To either side was a booster rocket, each 149 feet tall. Together, fuel tank and booster rockets held more than 4 million pounds of fuel. About two-thirds of it would be used in the first two minutes of the mission. The launch used a lot of fuel, but *Columbia* needed it to get into orbit.

9

The star of the show looked almost too **fragile** to withstand the vibration that 4 million pounds of fuel was going to kick off. Compared to the tank and boosters, the shuttle was small—120 feet long, with a **wing-span** of 80 feet. Clearly visible against the taller, darker forms, the shuttle had a graceful, eager look. Perhaps it was the swept-back delta wings. Perhaps it was the launch position, nose to the sky. Perhaps it was tired of waiting. Whatever the reason, the shuttle looked ready to go.

10

Millions of people around the world were also ready for the shuttle to go. Less than 20 years before, America and the world had watched Alan Shepard blast off. Space technology had come a long way: Shepard's Mercury capsule could fit inside the shell of one of the shuttle's main rocket engines.

11

Columbia's performance was spectacular. At 7:00 A.M. on April 12, 1981, the three huge main rocket engines rumbled awake. The rumble became a roar, as the two booster rockets also fired. **Billows** of smoke filled the launch pad. For one split-second, *Columbia* disappeared in its own exhaust. Then, steadily and surely, it climbed above the white billows. The shuttle looked as though it were being pushed by a wide column of smoke.

12

Just as they had watched 20 years before, Americans watched for a moment in awe and silence. Then, as they had 20 years before, Americans cheered and clapped and wept. *Columbia* continued its performance, climbing higher and higher toward orbit.

13

C. Finding the Main Idea

Highlight or circle the main idea in each paragraph. Remember that the main idea of a paragraph is often the first or last sentence in that paragraph. Other times, the main idea has to be pieced together from more than one sentence.

D. Making a Timeline

Place two important dates from this lesson on the timeline. Also place one important date from Lesson 8. Write two or three words to identify the importance of each date.

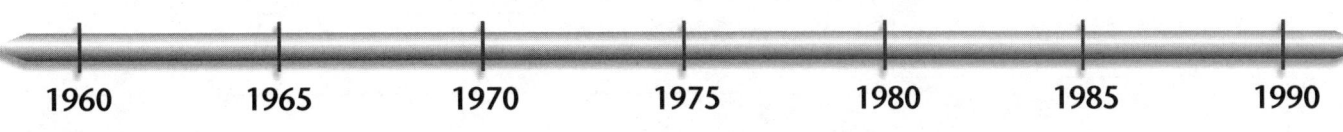

| 1960 | 1965 | 1970 | 1975 | 1980 | 1985 | 1990 |

E. Using Context Clues

Find the following four vocabulary words in the lesson. The words and sentences around each word give a clue to its meaning. Use these clues to write a meaning for each word. For each vocabulary word, describe one clue that you used.

1. fragile

Meaning: _____

_____

_____

Clue: _____

_____

2. engulfing

Meaning: _____

_____

_____

Clue: _____

_____

3. wing-span

Meaning: _____

_____

_____

Clue: _____

_____

4. billows

Meaning: _____

_____

_____

Clue: _____

_____

F. Reading for Details

Reread the paragraphs identified below. Then, complete the main idea/detail webs.

Paragraphs # 1–2 Main Idea: _____ _____ _____	Supporting Detail #1
	Supporting Detail #2
	Supporting Detail #3

Paragraph # 3 Main Idea: _____ _____ _____	Supporting Detail #1
	Supporting Detail #2
	Supporting Detail #3

G. Outlining the Text

The main idea/detail webs used in the Reading for Details section can be expanded and used to outline an entire chapter or lesson. Sometimes, small paragraphs can be grouped together on one web. Continuing from the maps on page 60, map the rest of the article. The main ideas are provided for you.

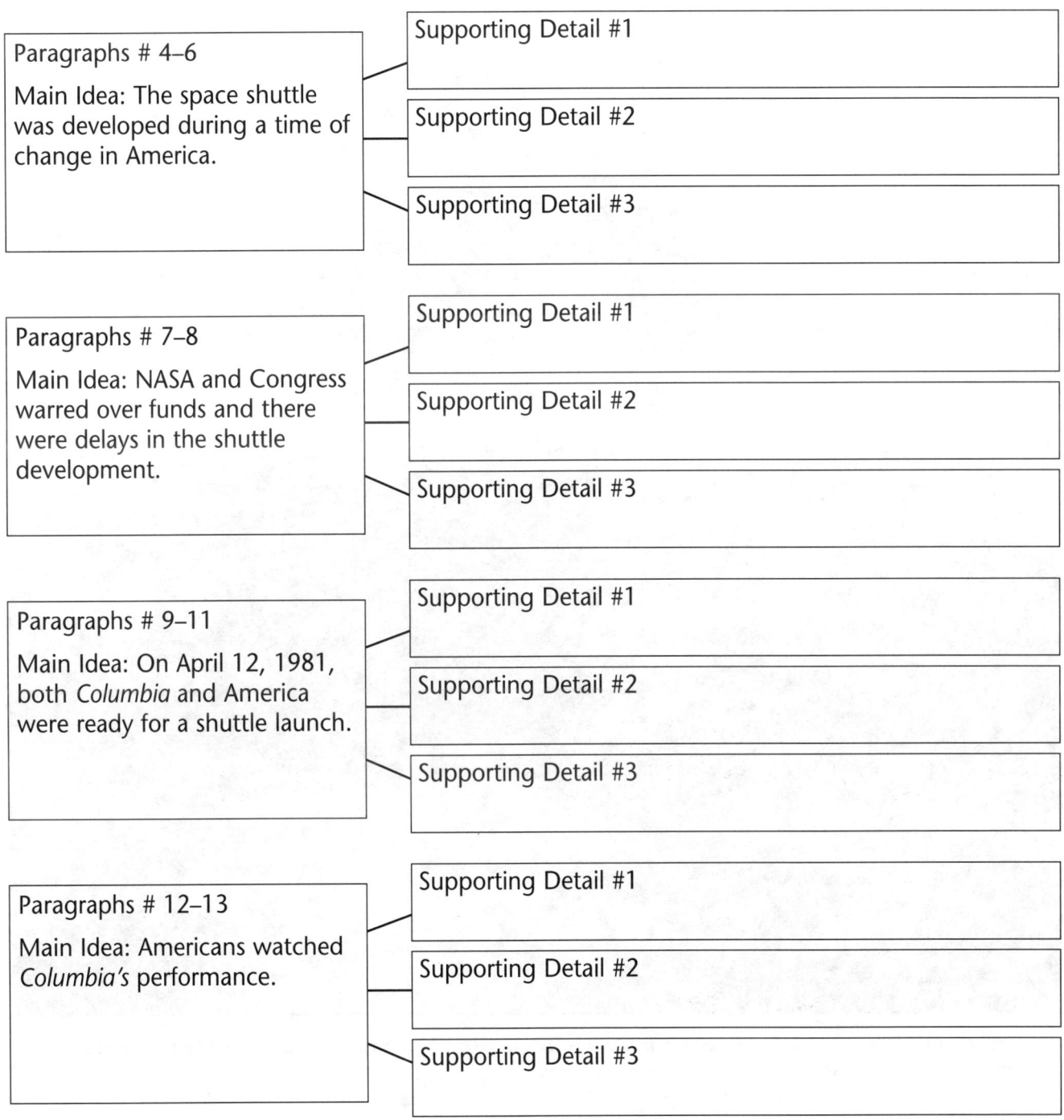

Paragraphs # 4–6

Main Idea: The space shuttle was developed during a time of change in America.

Supporting Detail #1

Supporting Detail #2

Supporting Detail #3

Paragraphs # 7–8

Main Idea: NASA and Congress warred over funds and there were delays in the shuttle development.

Supporting Detail #1

Supporting Detail #2

Supporting Detail #3

Paragraphs # 9–11

Main Idea: On April 12, 1981, both *Columbia* and America were ready for a shuttle launch.

Supporting Detail #1

Supporting Detail #2

Supporting Detail #3

Paragraphs # 12–13

Main Idea: Americans watched *Columbia's* performance.

Supporting Detail #1

Supporting Detail #2

Supporting Detail #3

Astronaut candidate Anna L. Fisher trains in zero-gravity. Astronauts must prepare for the weightless environment of the space shuttle.

The Space Shuttle Inside and Out

A. Setting the Stage

Read the first two and last two paragraphs of this lesson. Write four nouns or phrases that appear to be key ideas.

1.	2.	3.	4.

B. Discussing the Background

Reread the last two paragraphs of Lesson 9. As a group, discuss the relationship between Lesson 9 and Lesson 10.

Words to Know

heritage

monitor

galley

premoistened

suction

wired

modular

ascent

1 The shuttle was a new spacecraft, but its **heritage** could easily be seen. Features from other crafts were included both inside and out. Like an airplane, its flight deck was high on the nose. Also from the airplane, the shuttle borrowed wide windows and a pair of seats for the commander and copilot. The control panel looked like that of an Apollo spacecraft. However, the shuttle had many more controls than either an airplane or an Apollo. More than 2,000 switches, buttons, and levers controlled the shuttle's every operation and system.

2 To watch every button, switch, and lever would have been impossible. Instead, five computers **monitored** all equipment. If the computers detected a problem, the screens displayed all necessary information.

3 Behind the flight crew's seats were two more computers for mission specialists. These crew members operated the shuttle's equipment. They also launched or repaired satellites and performed other mission tasks. At the rear of the flight deck was a small area with windows that looked out over the shuttle's cargo bay. Underneath the windows were controls that worked the cargo bay's equipment.

4 In the left corner of the flight deck was a hatch (door) and ladder. They led to the mid-deck where the crew lived. The crew's quarters were much like those in *Skylab.* The quarters had a compact 2-foot wide kitchen, sleeping stations, storage lockers, a bathroom, and exercise equipment.

The shuttle offered all the comforts of home plus one: weightlessness. Every piece of equipment was designed so it would not float off if it was not stored. Some items, such as silverware, were magnetic. Others had Velcro patches.

5

Food was canned, dried, or frozen. All containers were capped. To mix food or add water, crew members poked a hole in the container. They drank liquids through straws. Frozen meals were heated in the **galley's** oven and served in the same tray.

6

> "You can't believe what a flying machine this is."
> —*Columbia* commander John Young, April 1981

In the sleeping compartments, the strap-in bags of *Skylab* had been replaced with zippered or Velcro-stripped bags. These bags could be attached to a bunk-like frame or to the wall. Since there is no "up" or "down" in space, the bed-bags could be placed at any angle.

7

For bathing, *Skylab* astronauts used a collapsible shower and **premoistened** towels. There was no sink. In space, water rises, forms little balls, and sticks to surfaces. The toilet had no water, either. Air flow **suctioned** out waste material, sending it to waste storage tanks.

8

The shuttle had a room-temperature atmosphere much like Earth's. Such an atmosphere was made possible by years of research. Previous astronauts had been **wired** with sensors that recorded the body's reactions in space. Scientists used this data to design and improve shuttle equipment. An important addition was exercise equipment and a workout schedule. This addition helped to keep the crew member's hearts and other muscles strong while in weightlessness.

9

Weightlessness, however, was one of the minor problems of space travel. More serious problems included possible radiation, cosmic rays, and extreme hot and cold temperatures. The crew was protected from these dangers by two devices: the shuttle's remarkable outer shield, and the shuttle program's space suit.

10

The shuttle's outer covering was made of more than 30,000 special ceramic tiles. Computers accurately measured and cut each tile. Then, the tiles were glued onto the shuttle by hand. Each tile was a different shape and thickness and had its own specific place on the craft. The tiles protected the shuttle's interior from extreme heat and cold. The tiles also kept the shuttle's aluminum frame from melting during reentry.

11

To save money, the shuttle program replaced the custom-fitted space suits of the moon projects. Money-saving **modular** suits were put together from about 20 mix-and-match parts. Astronauts could share the parts of the new suits. Many layers of nylon and Teflon were used. These materials were strong enough to resist puncture by space particles. The suits were also equipped with oxygen, air pressure, and temperature control systems.

12

When *Columbia's* engines roared to life on April 12, 1981, the noise and smoke was apparently all on the outside. The flight crew—astronauts John Young and Robert Crippen—described the shuttle's **ascent** as a ride in a "glass elevator." After two days of tests, Young danced a jig with thumbs up. "You can't believe what a flying machine this is," he reported.

13

NASA, however, put *Columbia* through its paces for three more test flights. At the end of its fourth test flight in 1982, NASA proclaimed it spaceworthy, and put it to work. *Columbia* began its working career on November 11, 1982. A four-person crew and two satellites blasted into orbit.

14

C. Finding the Main Idea

Highlight or circle the main idea in each paragraph. Remember that the main idea of a paragraph is often the first or last sentence in that paragraph. Other times, the main idea has to be pieced together from more than one sentence.

D. Making a Timeline

Place two important dates from this lesson on the timeline. Also place one important date from Lesson 9. Write two or three words to identify the importance of each date.

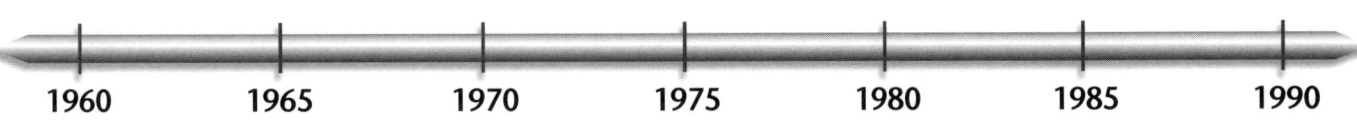

| 1960 | 1965 | 1970 | 1975 | 1980 | 1985 | 1990 |

E. Using Context Clues

Find the following four vocabulary words in the lesson. The words and sentences around each word give a clue to its meaning. Use these clues to write a meaning for each word. For each vocabulary word, describe one clue that you used.

1. heritage

Meaning: _____

_____

_____

Clue: _____

_____

2. galley

Meaning: _____

_____

_____

Clue: _____

_____

3. modular

Meaning: _____

_____

_____

Clue: _____

_____

4. ascent

Meaning: _____

_____

_____

Clue: _____

_____

F. Outlining the Text

Reread the following paragraphs. Then, complete the maid ideal/detail webs. The webs can be used to outline the entire lesson. Some of the main ideas are provided.

Paragraph # 1

Main Idea: _____

_____

_____

Supporting Detail #1

Supporting Detail #2

Supporting Detail #3

The Space Shuttle Inside and Out

Paragraphs # 2–3

Main Idea: Computers monitored all equipment.

Supporting Detail #1

Supporting Detail #2

Supporting Detail #3

Paragraphs # 4–8

Main Idea: _____

_____

_____

Supporting Detail #1

Supporting Detail #2

Supporting Detail #3

Paragraphs # 9–11

Main Idea: The crew was comfortable and protected.

Supporting Detail #1

Supporting Detail #2

Supporting Detail #3

Paragraph # 12

Main Idea: _____

_____

_____

Supporting Detail #1

Supporting Detail #2

Supporting Detail #3

Paragraphs # 13–14

Main Idea: *Columbia* proved to be a great flying machine.

Supporting Detail #1

Supporting Detail #2

Supporting Detail #3

Mission specialists George P. Nelson and James D. van Hoffen use *Challenger's* robot arm to help them repair the satellite *Solar Max*.

The Shuttle at Work

A. Setting the Stage

Read the first two and last two paragraphs of this lesson. Write four nouns or phrases that appear to be key ideas.

1.	2.	3.	4.

B. Discussing the Background

Reread the last two paragraphs of Lesson 10. As a group, discuss the relationship between Lesson 10 and Lesson 11.

Words to Know
punctual
dependable
adjustment
deploy
retrieve
versatility
tether

1 As a new NASA employee, *Columbia* had to prove itself. The shuttle had to show it was **punctual, dependable,** and could carry a fair work load. To do this, it followed a flight plan that never varied.

2 At T-minus two hours (two hours before lift-off), the crew boarded and began a final systems check. At T-minus nine minutes, the shuttle's computer team took over. At T-minus six seconds, the huge engines came alive. At T-minus zero, the rocket boosters fired. The whole system rose from the launch pad.

3 Two minutes after the shuttle had launched, the rocket boosters pushed it 28 miles above Earth. The rocket boosters' fuel was gone. They separated from the fuel tank and fell into the water for pick-up. Lightened, the shuttle and fuel tank climbed higher and went faster. At the same time, the shuttle drank the fuel tank dry. Eight minutes after liftoff, the shuttle arrived at the edge of space with an empty fuel tank. The fuel tank also separated from the shuttle and fell back to Earth. As it fell, the tank completely burned in the atmosphere.

4 The shuttle, freed from all its heavy tanks, approached orbit. The computers cut the three main engines and let *Columbia* coast for a few minutes. Then, two small rocket engines fired, boosting the shuttle into orbit. Soon after, the cargo bay doors opened to release the heat built up during ascent. The crew members unstrapped and began their duties.

When those duties were **accomplished,** the shuttle left orbit and reentered Earth's atmosphere. The astronauts switched on the two small rocket engines that had pushed the shuttle into orbit. The smaller rockets fired and the shuttle slowed down and dropped from orbit. The computers began to make **adjustments** that turned the shuttle from an orbiting spacecraft to a huge glider.

5

Spiraling down, the shuttle both slowed down and took the proper position for landing. At just the right moment, the computers brought down the landing wheels. The shuttle hit its home runway at 200 miles per hour.

6

The shuttle's procedure for getting into and out of orbit never varied. However, the tasks scheduled for each mission changed. Usually, the shuttle carried one or more satellites for the crew to **deploy.** Crew members might also have to **retrieve** a nonworking satellite from orbit and repair it in the cargo bay. Each crew also tested new equipment and new routines. To accomplish their tasks, crew members often used one of three pieces of space equipment. Each had been designed specially for the shuttle: the robot arm, *Spacelab,* and the Manned Maneuvering Units (MMUs).

7

"[From 1981–1985] we logged more time and flew more than twice as many individuals than in all previous history of human space flight in the United States."
—James Beggs, NASA Administrator, 1981–1985

The shuttle's arm was a 50-foot long robot built in Canada. The arm had human-like wrist, elbow, and shoulder joints. Mission specialists operated the arm from the rear flight deck inside the shuttle. The arm had its own lighting system and TV cameras. These tools let the crew members see their work up close. During NASA's first 24 shuttle missions, the arm lifted, moved, retrieved, or deployed more than 20 satellites.

8

Spacelab also added to the shuttle's **versatility.** The European Space Agency built the scientific workshop specifically for the shuttle. Spacelab had two sections that could be used together or separately. Inside, as many as four scientists could work in a comfortable environment. Scientists could perform advanced physics experiments, or observe the earth, sun or stars. An outside platform provided additional work space.

9

The third kind of equipment that contributed to the shuttle's usefulness were the Manned Maneuvering Units. These were jet-controlled backpacks that looked very much like armchairs. Wearing an MMU, an astronaut could fly free of the shuttle. **Tethers** such as those used by Apollo astronauts were not necessary. Astronauts could use units when doing external repairs or when building space structures, such as a space station.

10

By 1986, the shuttle had become the shuttles. NASA how had four sleek birds: *Columbia, Challenger, Discovery,* and *Atlantis.* The four crafts and their crews had flown 24 successful missions. Companies and governments were buying space on the shuttles. Scientists from other countries were passengers. So was a United States Senator. The shuttles were definitely a part of everyday life.

11

Shuttle missions had become so regular that people were no longer interested. The 25th shuttle mission excited people, however. An ordinary American citizen, teacher Christa McAuliffe, was going to space. From the shuttle, she planned to teach two classes to America's school children.

12

But *Challenger* taught America quite a different lesson from the one McAuliffe had planned. On January 28, 1986, 73 seconds after launch, *Challenger* exploded. McAuliffe and all of her six crew mates died. America learned that space travel was not yet an everyday affair. The explosion reminded Americans that space travel is dangerous.

13

C. Finding the Main Idea

Highlight or circle the main idea in each paragraph. Remember that the main idea of a paragraph is often the first or last sentence in that paragraph. Other times, the main idea has to be pieced together from more than one sentence.

D. Making a Timeline

Place one important date from this lesson on the timeline. Also place two important dates from Lesson 10. Write two or three words to identify the importance of each date.

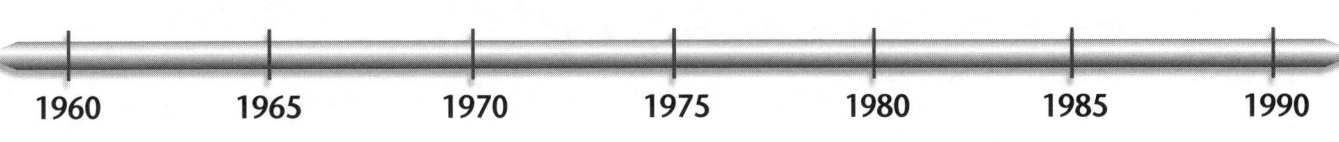

| 1960 | 1965 | 1970 | 1975 | 1980 | 1985 | 1990 |

E. Using Context Clues

Find the following four vocabulary words in the lesson. The words and sentences around each word give a clue to its meaning. Use these clues to write a meaning for each word. For each vocabulary word, describe one clue that you used.

1. adjustments

Meaning: _____

_____

_____

Clue: _____

_____

2. deploy

Meaning: _____

_____

_____

Clue: _____

_____

3. versatility

Meaning: _____

_____

_____

Clue: _____

_____

4. tethers

Meaning: _____

_____

_____

Clue: _____

_____

F. Outlining the Text

Reread the following paragraphs. Then, complete the main idea/detail webs.

Paragraphs # 1–6 Main Idea: _____ _____ _____	Supporting Detail #1
	Supporting Detail #2
	Supporting Detail #3

Paragraph # 13 Main Idea: _____ _____ _____	Supporting Detail #1
	Supporting Detail #2
	Supporting Detail #3

Understanding Complex Sentences

In this unit, you will focus on complex sentences. Complex sentences are word groupings that can be confusing or difficult to read. They often fit into one or more of these categories:

1. Sentences with many words.

2. Sentences with punctuation marks other than periods.

3. Sentences with unusual word combinations.

The following strategies will help you to successfully read complex sentences.

1. When a sentence has many words (more than 20), break the sentence into parts. Figure out what each part means and then put the meanings together.

2. Use punctuation marks as clues. For example:

Comma (,)—A comma is a sign to pause or a way to separate related parts.

Dash (–)—A dash is a way to set aside part of a sentence. Make sure you understand the information in front of a dash before continuing to read after the dash.

Colon (:)—A colon is often a sign that a definition or a list will follow.

Semicolon (;)—A semicolon is a sign that a sentence has two related thoughts. Read and understand the first thought before reading the second thought.

3. If a sentence has an unusual combination of words, break the sentence into smaller parts and find meaning for each part. Then, put the smaller parts together.

Space shuttle *Challenger* blasts off on June 18, 1983, for its second mission.

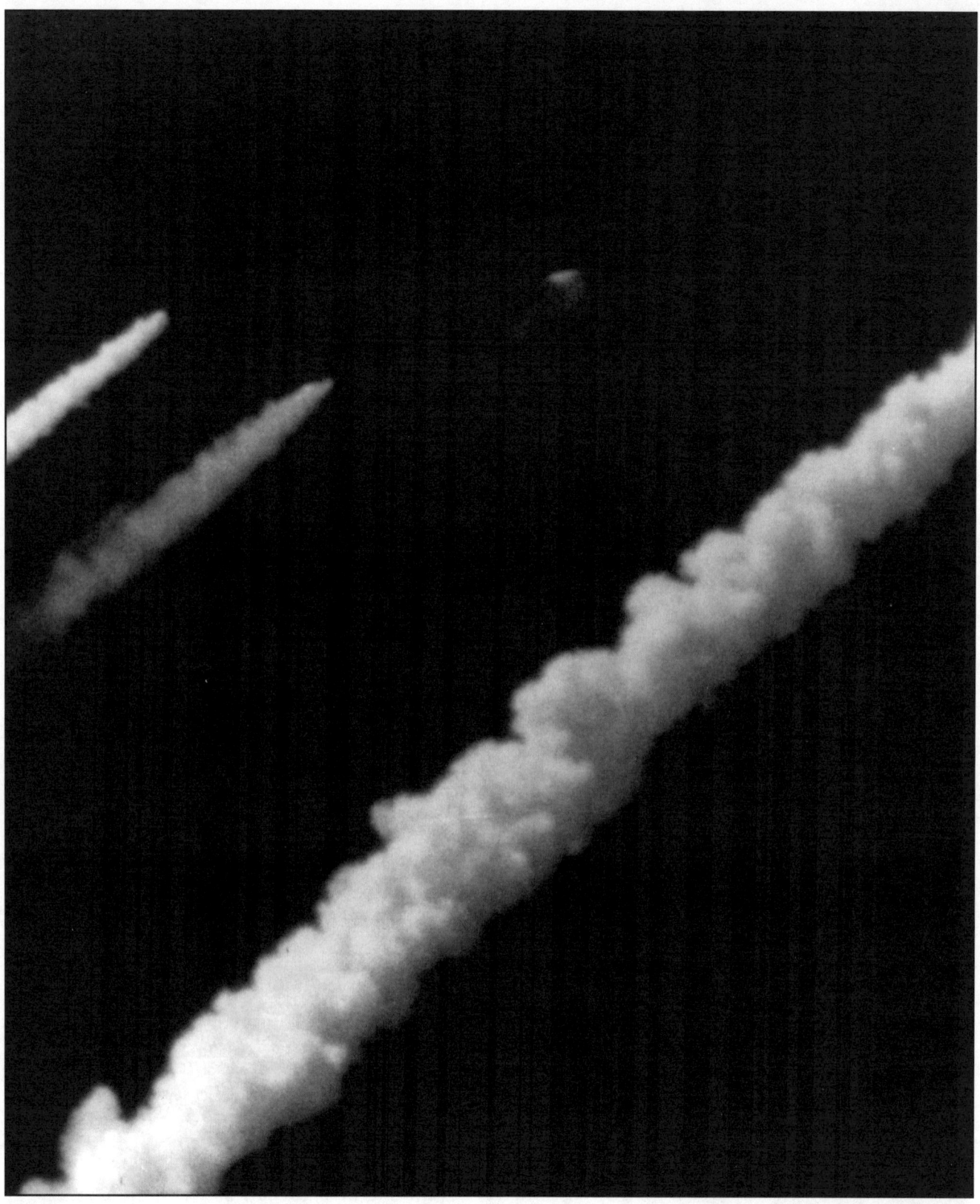

Plumes of smoke were the only signs of the *Challenger* shuttle after it exploded 73 seconds from launch on January 28, 1986.

A. Setting the Stage

Read the first two and last two paragraphs of this lesson. Write four nouns or phrases that appear to be key ideas.

1.	2.	3.	4.

B. Discussing the Background

Reread the last two paragraphs of Lesson 11. As a group, discuss the relationship between Lesson 11 and Lesson 12.

Words to Know

expand

watchdog agency

recommendation

shrouded

anticipation

apprehension

1 Of the four shuttles, *Challenger* was the true workhorse. It flew regularly and dependably, completing nine missions since its appearance on April 4, 1983. No one was prepared for the scene on national television on January 28, 1986.

2 *Challenger's* launch looked normal—for 73 seconds. First came the shuddering roar of the engines, then the froth of smoke out of which rose the *Challenger* clinging to its fuel-launch system. The familiar but still impressive column of smoke, for the tenth time, pushed the machine toward space. Suddenly, the column knotted up. A huge ball of smoke appeared where the *Challenger* should have been. Then a loop became visible, and a trail of exhaust headed straight down for the Atlantic Ocean.

3 Grief and disbelief stunned America. This was the first in-flight tragedy in America's space program. In-school televisions allowed the country's children to watch the tragedy. Twenty-five years had passed since the fire on *Apollo 1*. People had come to believe that NASA was totally in control.

4 President Ronald Reagan formed a 12-person investigation committee. Among its members were some well-known names: Chuck Yeager, Neil Armstrong, and Sally Ride. For four months, the members read reports, watched playbacks of the disaster, and interviewed people who had been involved. At the end of the four-month inquiry, the committee published its findings.

Challenger's explosion appeared to be caused by the malfunction of a rubber O-ring. This part was located between two sections of the right booster rocket. During the launch, the O-ring was supposed to **expand** and fill the space between the sections. But temperatures at Cape Canaveral were very cold. Since the O-ring was cold, it did not expand to fill up the space. Fuel leaked out, burned through the wall of the huge fuel tank, and ignited.

5

Temperatures at the Cape the morning of *Challenger's* launch had dropped to the low 20s. These were unusually low temperatures for Cape Canaveral. NASA had not done testing with such coolness. The scientists did not realize they had a problem. Going ahead with the shuttle launch under untested conditions proved to be a horrible mistake.

6

"Every family member I talked to asked specifically that we continue the program . . . . We will not disappoint them."
—Ronald Reagan, U.S. President, *Challenger* Eulogy, January 31, 1986

Learning that the disaster could have been avoided angered a saddened Congress and America. Should NASA be "punished," put under a **watchdog agency** or committee? Should NASA be closed down altogether? Should the shuttle program be scrapped?

7

The committee's answer was a resounding "No!" The country now knew NASA was not perfect. However, the committee believed NASA was well equipped to avoid another accident. The commission gave NASA nine **recommendations** for improvement. The first was, of course, to redesign the O-rings. The second was to form a safety committee to develop a system of launch checks.

8

The *Challenger* tragedy wounded both NASA and America. Space travel was still dangerous. Spaceflight was not an everyday part of life, and might not be for many years.

9

For months after the *Challenger* tragedy, NASA worked quietly. The organization needed time to reorganize and heal. The agency's management process was changed, the safety panel was created, and new procedures were put into place. America, too, was healing. Desire to honor *Challenger's* crew slowly replaced the nation's grief and disappointment. People began to talk about another shuttle launch.

10

The *Challenger* Tragedy

On September 29, 1988, *Discovery* sat on the launch pad. New people, parts, and procedures—*Discovery* represented almost a whole new space program. There was a new feeling in the air, too: a mixture of **anticipation** and **apprehension.** Everyone knew what could happen.

11

At 11:37 A.M. came a sound that had not been heard in more than two years. *Discovery's* three main engines rumbled the opening words of a new chapter in space flight. The two rocket boosters quickly added their roars. *Discovery* was **shrouded** in the foam of blast-off. Millions of Americans watched in silence, remembering a similar scene 32 months before.

12

Discovery rose on its column of smoke and fire, heading for space. Twenty seconds, 40, 60, 73: "We have lift-off, we have lift-off." *Discovery's* roar was joined by the roar of the crowd.

13

By early 1989, *Discovery* and sister ship *Atlantis* were cutting into the backlog of jobs that had built up in the months following the explosion of *Challenger.* A new space shuttle was in the works.

14

So was a new dream. In 1988, the United States signed a space station agreement. On behalf of the U.S., NASA began working with other nations to build an International Space Station (ISS). The first phase of the ISS has been in orbit since the fall of 1998.

15

C. Finding the Main Idea

Highlight or circle the main idea in each paragraph. Remember that the main idea of a paragraph is often the first or last sentence in that paragraph. Other times, the main idea has to be pieced together from more than one sentence.

D. Making a Timeline

Place three important dates from this lesson on the timeline. Also place one important date from Lesson 11. Write two or three words to identify the importance of each date.

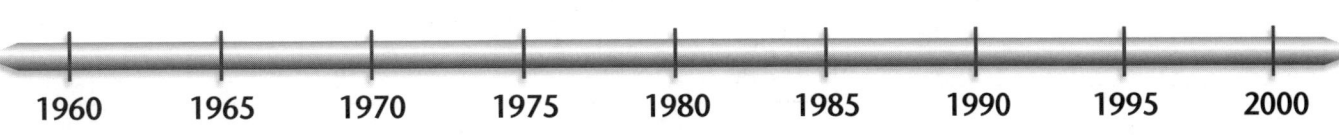

1960   1965   1970   1975   1980   1985   1990   1995   2000

E. Using Context Clues

Find the following four vocabulary words in the lesson. The words and sentences around each word give a clue to its meaning. Use these clues to write a meaning for each word. For each vocabulary word, describe one clue that you used.

1. expand

Meaning: _____

_____

_____

Clue: _____

_____

2. recommendations

Meaning: _____

_____

_____

Clue: _____

_____

3. apprehension

Meaning: _____

_____

_____

Clue: _____

_____

4. shrouded

Meaning: _____

_____

_____

Clue: _____

_____

F. Reading for Details

Reread the paragraphs identified below. Then, complete the main idea/detail webs.

Paragraphs # 5–6 Main Idea: _____	Supporting Detail #1
	Supporting Detail #2
	Supporting Detail #3

Paragraphs # 10–15 Main Idea: _____	Supporting Detail #1
	Supporting Detail #2
	Supporting Detail #3

G. Understanding Complex Sentences

Four complex sentences are listed below. Match the sentences with their descriptions by writing letters on the blanks. Then, write the sentence parts as directed.

1. ___ There was a new feeling in the air, too: a mixture of anticipation and apprehension.

a. Commas are used to separate parts of a list. Rewrite the list using bullets.

-

-

-

2. ___ For four months, the members read reports, watched playbacks of the disaster, and interviewed people who had been involved.

b. A colon is used to introduce an explanation of the first part of the sentence. Write the explanation.

3. ___ New people, new parts, new procedures—*Discovery* represented almost a whole new space program.

c. Commas are used to separate three small sentences that have been combined into one long sentence. Write the middle sentence.

4. ___ The agency's management process was changed, the safety panel was created, and new procedures were put into place.

d. Commas are used to separate three details that support the part of the sentence that follows the dash. Use bullets to list the three details.

-

-

-

Space shuttle *Endeavor* docks with the international space station *Freedom*.

The Future

A. Setting the Stage

Read the first two and last two paragraphs of this lesson. Write four nouns or phrases that appear to be key ideas.

1.	2.	3.	4.

B. Discussing the Background

Reread the last two paragraphs of Lesson 12. As a group, discuss the relationship between Lesson 12 and Lesson 13.

Words to Know

Cold War

assemble

submit

fleet

vital

debris

1 The 1990s were a time of great developments for NASA and international space interests. Three programs were started that have set the course for the future of space exploration: the International Space Station (ISS), the Missions to Mars, and the Hubble Telescope.

2 The dream of international cooperation in space has come true with the building of the ISS. In 1988, the United States, Canada, Japan, and the European Space Agency (Belgium, Denmark, France, Germany, Great Britain, Italy, The Netherlands, Norway, Spain, Sweden, and Switzerland) were the initial partners in the station's planning. Following the end of the **Cold War,** discussions with Russia began. In 1993, Russia officially joined the group.

3 Although the Russians came into the ISS planning late, they had much to offer. Throughout the 1990s, the Russian space station, *Mir,* orbited the earth. Beginning in 1995, U.S. staff were stationed on *Mir* with Russian cosmonauts and scientists. Experiments on *Mir* provided insight into the needs of international work groups. Also, in many ways, *Mir* served as a guide in the design of the ISS. People spent months at a time on *Mir* as opposed to a maximum of two weeks in a NASA shuttle or a couple of months with *Skylab.* Although NASA had **logged** more than 800 days in shuttles, *Mir* still provided the most long-term stay data and the only international data.

The first phase of the ISS was **assembled** in Earth's orbit in the fall of 1998. From the beginning, the U.S. shuttles were key figures in the construction of the station. The ISS plans call for a permanent laboratory, facilities for docking shuttles, and an orbital service center. Countries not involved in the partnership will be able to buy space and time on the shuttle and the station.

4

"It's time for the human race to enter the solar system."
—Dan Quayle, U.S. Vice President, 1989–1993

Scientists are making detailed plans for the ISS laboratory. Many of the great inventions of the 21st century will likely be due, at least in part, to the ISS laboratory work. The lack of gravity in space is one of the conditions that makes space research useful. Some experiments that gravity interferes with on Earth can easily be performed in space.

5

In 1989, President Bush attended the 20th anniversary celebration of the moon landing. He called for the U.S. to land a person on Mars. People became excited and NASA started to assemble a budget. The first Mars plan **submitted** by NASA was so expensive that talk of going to Mars cooled. Then NASA offered a new plan and Mars missions started.

6

NASA has had successes and failures with the Mars project. *Observer* disappeared in 1993, just three days before it was to go into Mars' orbit. On July 4, 1997, the Pathfinder Rover named *Sojourner* landed on Mars. Pictures and test results were exciting: perhaps living creatures had once survived on Mars! The combination of the successful mission and the new possibilities increased the desire to head to Mars. In 1999, within four months of each other, *Climate Orbiter* and *Polar Lander* failed to reach Mars and were no longer heard from. Luckily, before they were lost, both were able to return new data to Earth. *Global Surveyor* entered Mars's orbit on September 12, 1997. Along with sending information to Earth, *Surveyor* looked for signs of *Observer* and found nothing. *Surveyor* was still in Mars' orbit in the year 2000.

7

In 1990, the Hubble Telescope was placed into orbit around the earth. This eye-in-space provides new insight into the stars, earth, moon, sun, and other bodies in space. At 43 feet long, Hubble is the largest earth item ever to go into orbit. It can "see" farther and clearer than any other telescope. With the information collected, scientists will come to better understand our solar system and beyond.

8

During the decade following the *Challenger* disaster, NASA successfully launched more than 70 shuttle missions. The **fleet** of shuttles included *Atlantis, Columbia, Discovery,* and *Endeavour.*

9

Shuttles continue to be **vital** to NASA's space program. The busy shuttles are helping to build the ISS, servicing the Hubble Telescope, launching and retrieving satellites, and doing science research. One *Discovery* mission created a lot of interest in the fall of 1998. Former astronaut turned U.S. Senator, 77-year-old John Glenn, returned to space for a 9-day mission. He helped with a study on space flight and the aging process.

10

A drawback to the space program appears to be that humans are littering Earth's orbit. Studies show that more than 60,000 pieces of space junk, ranging in width from inches up to feet, are orbiting Earth right now. About 5 percent of the pieces are operating spacecraft and 20 percent are satellites that no longer work. Many other pieces are discarded material from space operations. In addition to abusing the environment, this **debris** can interfere with satellite transmissions and astronaut work.

11

Where is space exploration headed? The possibilities are amazing. Humans may land on Mars and perhaps return to the moon. Also, space research may result in new findings that could change the world.

12

C. Finding the Main Idea

Highlight or circle the main idea in each paragraph. Remember that the main idea of a paragraph is often the first or last sentence in that paragraph. Other times, the main idea has to be pieced together from more than one sentence.

D. Making a Timeline

Place three important dates from this lesson on the timeline. Also place one important date from Lesson 12. Write two or three words to identify the importance of each date.

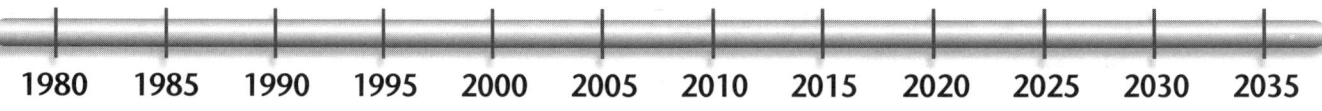

1980	1985	1990	1995	2000	2005	2010	2015	2020	2025	2030	2035

E. Using Context Clues

Find the following four vocabulary words in the lesson. The words and sentences around each word give a clue to its meaning. Use these clues to write a meaning for each word. For each vocabulary word, describe one clue that you used.

1. assembled

Meaning: _____

_____

_____

Clue: _____

_____

2. fleet

Meaning: _____

_____

_____

Clue: _____

_____

3. submitted

Meaning: _____

_____

_____

Clue: _____

_____

4. debris

Meaning: _____

_____

_____

Clue: _____

_____

F. Reading for Details

Reread the two paragraphs identified below. Then, complete the main idea/detail webs.

Paragraph # 1 Main Idea: _____ _____	Supporting Detail #1
	Supporting Detail #2
	Supporting Detail #3

Paragraph # 7 Main Idea: _____ _____	Supporting Detail #1
	Supporting Detail #2
	Supporting Detail #3

| G. Understanding Complex Sentences |

Four complex sentences are listed below. Match the sentences with their descriptions by writing letters on the blanks. Then, write the sentence parts as directed.

1. ___ The ISS plans call for a permanent laboratory, facilities for docking shuttles, and an orbital service center.

a. Commas are used to separate parts of a list. Rewrite the list using bullets:

•

•

•

2. ___ Studies show that more than 60,000 pieces of space junk, ranging in width from inches up to feet, are orbiting Earth right now.

b. Parentheses are used to provide some in-depth information. Write the sentence without the in-depth information.

3. ___ Three programs were started that have set the course for the future of space exploration: the International Space Station (ISS), Missions to Mars, and the Hubble Telescope.

c. Commas are used to separate the main sentence from an additional-information phrase. Write the main sentence.

4. ___ In 1988, the United States, Canada, Japan, and the European Space Agency (Belgium, Denmark, France, Germany, Great Britain, Italy, The Netherlands, Norway, Spain, Sweden, and Switzerland) were the initial partners in the station's planning.

d. A colon is used to introduce a list. Rewrite the list using bullets.

•

•

•

Light, long-lasting wheelchairs are made from *Lunar Rover* materials.

Many athletic shoes are cushioned by woven coils layered over the sole. The coil fabric was adapted from moon boots.

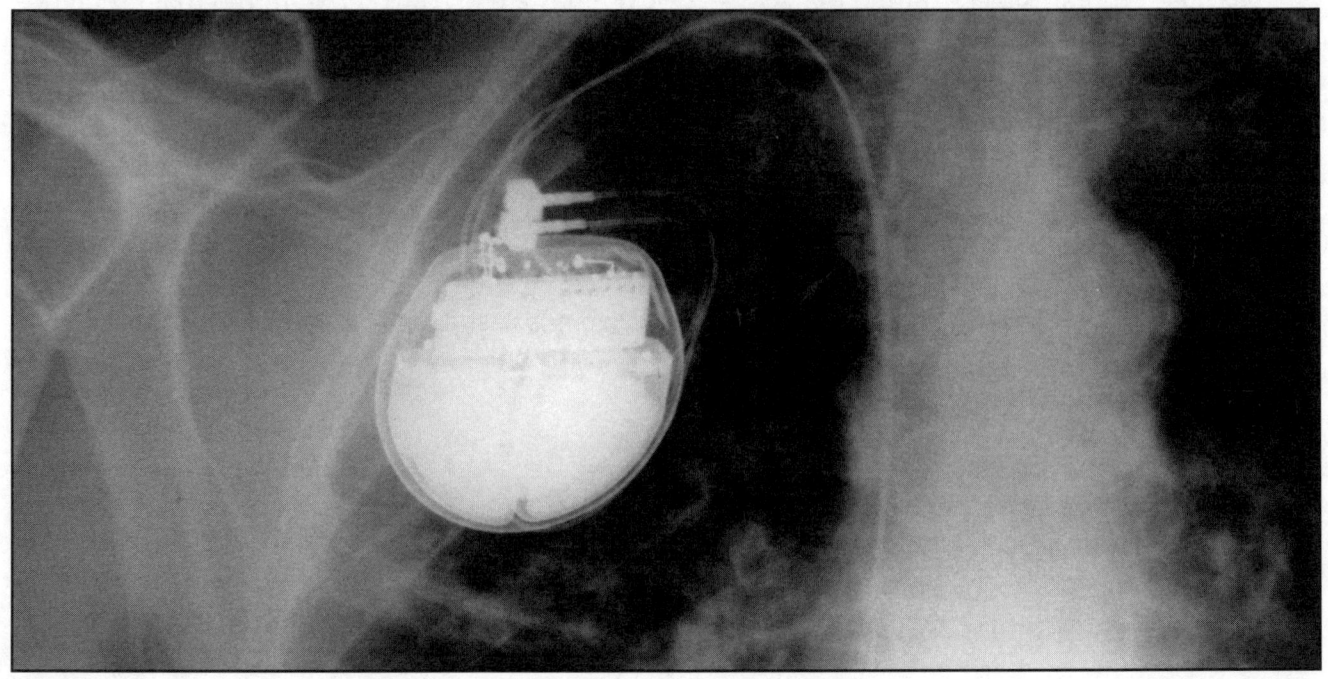

The space program led to the development of miniature equipment, like pacemakers. It also gave us digital image processing, which can help doctors better read X-rays.

A. Setting the Stage

Read the first two and last two paragraphs of this lesson. Write four nouns or phrases that appear to be key ideas.

1.	2.	3.	4.

B. Discussing the Background

Reread the last two paragraphs of Lesson 13. As a group, discuss the relationship between Lesson 13 and Lesson 14.

Words to Know
protective
implant
telemetry
transmit
emphysema
influenza
porous
benefit

1 On July 20, 1969, astronauts Neil Armstrong and Buzz Aldrin walked on the moon. They wore moon boots that had a special layer of plastic coils sewn next to the soles. These coils cushioned the astronauts' feet during their moon walk. Today, the coil material is used in many shoes worn by joggers and walkers. The coils make the shoes more comfortable and soak up the shock when the feet hit the ground.

2 The helmets worn by the two astronauts had special visors to protect their eyes from the sun's harmful rays. Today, millions of people who work or play in the sun wear sunglasses or goggles that have the same **protective** materials.

3 Armstrong and Aldrin also wore space suits made from special fire-resistant material. Today, that same material covers seats in most aircraft and is made into protective clothing for fire fighters and race car drivers.

4 The space program needed many small, tough, long-lasting items. This need led to the development of miniature tools and equipment that are now used in medicine. Surgeons can **implant** tiny pacemakers in patients with heart problems. The pacemakers give people a new chance at life. Credit-card sized medicine pumps measure and inject medicine directly into a person's bloodstream. Neither doctor nor patient need worry about proper doses and timing.

Materials for spacecraft and equipment have to be small, tough, light, and long-lasting. The *Lunar Rover* materials are now used to make light, long-life wheelchairs. These chairs weigh only 25 pounds and can be handled easily by those who use them.

5

The space program also provides research beyond its own needs. More and more, research is being conducted outside of Earth's gravity. As the following examples show, the results are amazing!

6

". . . we have had many amazing accomplishments in the space program recently. The Space Shuttle has been a platform for incredible experiments and amazing discoveries."
—Daniel S. Goldin, NASA Administrator, 1992–present

Space-grown crystals develop more perfectly due to the lack of gravity. These crystals have been used in making new drugs for AIDS, **emphysema**, **influenza**, and diabetes.

7

Experiments in space beginning in 1994 resulted in an understanding of how metal is formed. This knowledge has led to the ability to make better metals.

8

The space program's missions have given us a new understanding of how fire works. This information has helped both in controlling fires and in using fire power.

9

Scientists have learned that contact lenses prepared in space are more uniform and **porous**. These lenses allow more oxygen to pass to the eyes.

10

Cancer research is getting a huge boost from gravity-free research. The low gravity state allows scientists to grow samples of cancers. Scientists can then test treatments with no risk to a patient.

11

The space program uses **telemetry** to send data from space to a computer on Earth. The computer decodes the data and makes a picture. Doctors use the same procedure with people who have muscle problems. Sensors are attached to both a muscle and a computer. As the patient moves the muscle, coded data is sent to the computer. The computer creates a picture that helps doctors pinpoint problems.

12

Doctors are also using another space communication process—digital image processing. With this process, doctors can better read X-rays, CAT scans, or other diagnostic pictures. It allows doctors to choose which body parts to include on an X-ray. In this way, doctors find and measure tumors or tears in tissue or muscles.

13

Since 1958, NASA has launched weather, communications, and scientific satellites on a regular basis. The returned data has improved weather forecasts. Weather satellites can also spot erupting volcanoes, blazing forest fires, and sources of air and water pollution.

14

Communications satellites send thousands of signals every day. From high above the earth, satellites **transmit** phone calls, television shows, and radio programs across towns, countries, and oceans. In seconds, these satellites send huge amounts of data across thousands of miles.

15

These examples are only some of more than 30,000 ways that people have gained from technology developed for and by the space program. Every American has, in some way, **benefited** from the program.

16

Douglas Morrow, a former member of the NASA Advisory Council, sums it up: "Everyone talks about the dawn of the Space Age. But the Space Age is already here. The sun is up and it shines on everyone."

17

C. Finding the Main Idea

Highlight or circle the main idea in each paragraph. Remember that the main idea of a paragraph is often the first or last sentence in that paragraph. Other times, the main idea has to be pieced together from more than one sentence.

D. Making a Timeline

Place two important dates from this article on the timeline. Also place one important date from Lesson 13. Write two or three words to identify the importance of each date.

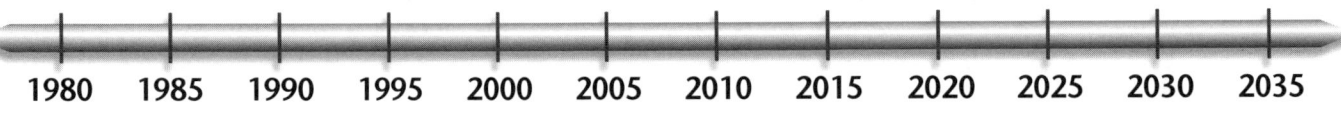

| 1980 | 1985 | 1990 | 1995 | 2000 | 2005 | 2010 | 2015 | 2020 | 2025 | 2030 | 2035 |

E. Using Context Clues

Find the following vocabulary words in the lesson. The words and sentences around each word give a clue to its meaning. Use these clues to write a meaning for each word. For each vocabulary word, describe one clue that you used.

1. implant

Meaning: _____

_____

_____

Clue: _____

_____

2. porous

Meaning: _____

_____

_____

Clue: _____

_____

F. Reading for Details

Reread the paragraph identified below. Then, complete the main idea/detail web.

Paragraph # 1

Main Idea: _____

_____

_____

Supporting Detail #1

Supporting Detail #2

Supporting Detail #3

G. Understanding Complex Sentences

Find the complex sentences described below. Follow the directions for each.

a. Find a sentence in Paragraph 13 that uses commas to separate parts of a list. Underline the sentence.

b. Find a sentence in Paragraph 16 that uses more than 20 words without any guiding punctuation. Underline the sentence.

c. Find a sentence in Paragraph 13 that uses a dash to separate a phrase from its definition. Underline the sentence.

d. Find a sentence in paragraph 15 that uses commas to separate parts of a list. Circle the commas.

End-of-Book Test

A. Write *True* **or** *False* **for each sentence.**

_____ 1. The first Apollo mission ended in disaster on the launch pad.

_____ 2. Space shuttles are reusable rocket planes.

_____ 3. The Lunar Rover was a dune-buggy-type vehicle used on the moon.

_____ 4. After the *Challenger* accident, NASA was shut down.

_____ 5. Both rockets and kites were invented by the Chinese.

_____ 6. *Skylab* was built after the U.S. had landed on the moon.

_____ 7. The Wright brothers invented the first liquid-fueled rocket.

_____ 8. Research done in outer space has helped people on Earth.

_____ 9. Sputnik was an electrical problem in one of the first shuttles.

_____ 10. Few Americans were watching the *Challenger* launch when the shuttle blew up.

B. Match each item on the left with the correct detail on the right.
Write each answer on the line.

_____ 1. reusability   **a.** outside the spacecraft

_____ 2. deploy   **b.** Russian astronaut

_____ 3. lunar   **c.** a planned, skillful movement

_____ 4. space shuttle   **d.** the most unique feature of the space shuttle

_____ 5. supersonic   **e.** a space project involving several countries

_____ 6. MMU   **f.** having to do with the moon

_____ 7. extravehicular   **g.** to release a satellite in space

_____ 8. ISS   **h.** allows astronauts to travel outside of the shuttle in space

_____ 9. maneuver   **i.** lands like an airplane

_____ 10. cosmonaut   **j.** faster than sound

C. Circle the word or phrase that correctly completes each sentence.

1. The first NASA shuttle was (*Discovery, Challenger, Columbia*).

2. The United States hopes to land a person on (Saturn, Venus, Mars) by 2020.

3. The first satellite launched into space was (*V-2, Sputnik, Gemini*).

4. The space shuttle's (delta wings, galley, tiles) made the shuttle easier to handle and made landing safer.

5. The first space shuttle was created by (the U.S., the Soviet Union, Germany).

6. *Mir* space station belongs to (the U.S., Russia, Germany).

7. When spacecraft meet in space, they (deploy, rendezvous, recap).

8. The first war-time rocket, the V-2, was launched by (the U.S., the Soviet Union, Germany).

9. The first person to walk on the moon was (Yuri Gagarin, Neil Armstrong, Christa McAuliffe).

10. The first person to orbit Earth was from (the U.S., the Soviet Union, Germany).

D. Choose the correct answer. Write each answer on the line.

1. U.S. Senator and former astronaut _____ returned to space in his seventies.
   a. Werner Von Braun   **b.** Buzz Aldrin   **c.** John Glenn   **d.** Neil Armstrong

2. A spacecraft's upward climb is called _____ .
   a. ascent   **b.** deployment   **c.** descent   **d.** retrieval

3. _____ built and launched the first liquid-fueled rocket in the U.S.
   a. Neil Armstrong   **b.** Robert Goddard   **c.** John Glenn   **d.** Werner Von Braun

4. The first person in space was _____ .
   a. John Glenn   **b.** Neil Armstrong   **c.** Alan Shepard   **d.** Yuri Gagarin

5. _____ was the teacher who died in the *Challenger* accident.
   a. Robert Goddard   **b.** Gus Grissom   **c.** Sally Ride   **d.** Christa McAuliffe

6. _____ was a plan to land an astronaut on the moon.
   a. Project Gemini   **b.** Operation Paperclip **c.** Project Apollo **d.** Project Mercury

7. A group of spacecraft is called a _____ .

   a. mass   **b.** fleet   **c.** tether   **d.** station

8. _____ send information from space to Earth.

   a. Thrusters   **b.** Satellites   **c.** Boosters   **d.** Galleys

9. Chuck Yeager's flight broke the _____ barrier.

   a. sound   **b.** light   **c.** space   **d.** lunar

10. Space suit material is now also used to make clothing for _____.

   a. farmers   **b.** fire fighters   **c.** doctors   **d.** joggers

E. Answer the following questions with complete sentences.

1. Why did NASA work to build a reusable spacecraft?

   _____

   _____

   _____

   _____

2. What might be some future events in space travel?

   _____

   _____

   _____

   _____

Glossary

A

accurate—exact, precise, and effective (15)

achieve—to accomplish or complete something (51)

adapt—to change to meet certain needs (10)

adjustment—a change or correction that makes something better (70)

advanced—highly developed or complicated (27)

annihilate—to completely destroy (21)

anticipation—the act of looking forward to something with excitement (77)

apprehension—the act of looking forward to something with fear (77)

ascent—upward climb (65)

assemble—to put together (82)

awe—amazement and wonder (47)

B

benefit—to receive positive help (89)

billow—a large rolling mass; a puff or surge of smoke (59)

bleak—not encouraging or hopeful (45)

C

Cold War—political and economic conflict between the United States and the Soviet Union after World War II (81)

complex—complicated; having many parts (53)

contraption—a device, gadget, or creation (9)

convert—to change (51)

convince—to cause someone to believe or feel sure (9)

cosmonaut—an astronaut from the Soviet Union (now Russia) (29)

D

data—information, often represented by numbers (16)

debris—pieces or fragments scattered about (83)

delta wings—aircraft wings in the shape of large triangles (53)

dependable—reliable or trustworthy (69)

deploy—to release (70)

device—something, such as a tool, invented for a particular use (15)

disengage—to detach, free, or loosen from (41)

dominate—to lead or control (27)

E

emphysema—disease of the lungs that makes breathing difficult (88)

engulf—to take over or completely cover (57)

evolve—to change gradually over time (15)

expand—to get bigger or take up more space (76)

extensive—long; complete and thorough (52)

extravehicular—outside a spacecraft (39)

F

fatality—death (41)

feature—quality or characteristic (57)

fine-tune—to make small adjustments to something so it works better (22)

fleet—group of aircraft, watercraft, or spacecraft under one command (83)

fragile—delicate and very easily broken (58)

frenzy—great excitement or panic (28)

function—the purpose of something; what something is used for (57)

G

galley—kitchen of an aircraft (64)

H

heritage—customs, skills, or features handed down from one generation to the next (63)

humdrum—common, everyday, boring (47)

I

implant—to insert or set one thing into another (87)

impress—to amaze or cause wonder (15)

influenza—disease caused by a virus that has the symptoms of a severe cold; the flu (88)

ingenuity—cleverness or great skill in planning or doing something (47)

initial—first (23)

L

lunar—relating to the moon (45)

M

malfunction—to not work or perform properly or as expected (47)

maneuver—a planned, skillful movement (40)

mass—the amount of space an object takes up (52)

meteoroid—piece of rock or metal traveling through space (34)

modular—made of several parts (65)

monitor—to listen to or watch something closely to observe and record changes (63)

O

orbit—the curved path of one object traveling around another object in space (28)

outfit—to supply with the necessary materials or articles (51)

P

patented—got the rights for (16)

porous—full of holes through which liquids or gases may pass (88)

premoistened—made wet beforehand (64)

propel—to drive ahead or force outward (40)

propellant—liquid fuel (16)

protective—designed to prevent injury or harm (87)

punctual—on time (69)

puncture—a hole pierced in something (34)

R

recap—to summarize or redo (4)

recommendation—advice or suggestion (76)

reentry—stage at which a spacecraft returns through Earth's atmosphere (34)

rendezvous—to meet or gather

respond—to answer or react (33)

restore—to bring back to former condition (28)

retrieve—to recover or bring back (70)

S

satellite—device launched into orbit to collect and send data (23)

shrouded—covered or hidden (77)

simulated—artificial or not real; created by humans (52)

site—place (45)

Glossary

sound barrier—point at which aircraft approach the speed of sound (23)

splashdown—stage at which a spacecraft plunges into Earth's ocean (47)

squabble—a fight or disagreement (28)

streamlined—smooth; offering little resistance (17)

submit—to put forth an idea (82)

suborbital—consisting of less than a full orbit (40)

substitution—a replacement; the use of one thing for another (15)

suction—to draw liquids or gases into a space by removing air from that space (64)

superior—better than most others (33)

supersonic—faster than the speed of sound (23)

T

technology—scientific knowledge to develop or improve things (10)

telemetry—use of measuring devices to transmit information to a distant receiver (88)

tether—a rope or chain used for fastening (71)

theorize—to put forth an idea or theory (16)

thrust—the pushing power of a rocket or jet engine caused by the burning of fuels (11)

transmit—to send (89)

V

versatility—ability to be used for many different things (70)

vital—very important; needed for survival or success (83)

W

watchdog agency—an organization that keeps close watch on others' activities (76)

wing-span—distance between the wing tips of an airplane or spacecraft (58)

wired—set up with wire for electricity to travel through (64)

witness—to see an event as it happens (11)